Carbon Capture Technology and Storage in Poland

The book describes the results of the AGaStor Project's social research carried out in northwestern Poland which recognizes the main social opportunities and barriers to the introduction of CCUS to society.

At the national and regional level, the social barriers of CCUS acceptance are ignorance, uncertainty and the lack of public debate about this technology. On the local scale, social acceptance depends on the needs and fears of the local community. Some of those fears – distrust and perception of CO_2 as a waste – could be recognized as universal dilemmas connected with CCUS. Although as knowledge about CCUS increases, a significant change in attitude can be observed. This change does not mean an acceptance of that technology. It is rather a shift towards more benign forms of resistance – from NIMBY to WIMBY. The book presents actual mechanisms of social reaction to UGS and CCUS investment, which will be a valuable contribution to managing social change in the context of green transition.

The issue of social acceptance of CCUS should be of interest to decision makers and practitioners who manage investment projects in the broadly understood green transformation on a daily basis, and also to postgraduate students, researchers, academics and lecturers in political studies, sociology or economics, and in engineering studies connected with energy transformation.

Dariusz Wojakowski is a sociologist, professor in the Department of Society and Technology Studies, and Vice-Dean for Cooperation and Students at the Faculty of Humanities of the AGH University of Kraków, Poland.

Oluf Langhelle is Professor in political science at the University of Stavanger, Department of Media and Social Sciences, Norway.

Mateusz Stopa, PhD, is a methodologist and project manager at SOLIDATA (research company, Poland).

Stanisław Nagy is a professor at the Department of Gas Engineering, AGH University of Krakow, Poland.

Barbara Gąciarz is a professor at the AGH University of Kraków, Poland, and has a PhD in sociology and economics.

Thomas Sattich is Associate Professor at the Department of Media and Social Science, University of Stavanger, Norway.

Routledge Focus on Energy Studies

Limits to Terrestrial Extraction
Edited by Robert E. Kirsch

Network Governance and Energy Transitions in European Cities
Timea Nochta

Residential Electricity Consumption in Urbanizing China
Time Use and Climate-Friendly Living
Pui Ting Wong and Yuan Xu

Carbon Capture Technology and Storage in Poland
Social Acceptance and the Energy Transition
Dariusz Wojakowski, Oluf Langhelle, Mateusz Stopa, Stanislaw Nagy, Barbara Gąciarz and Thomas Sattich

For more information about this series, please visit: www.routledge.com/Routledge-Focus-on-Energy-Studies/book-series/RFENS

Carbon Capture Technology and Storage in Poland

Social Acceptance and the Energy Transition

Dariusz Wojakowski, Oluf Langhelle, Mateusz Stopa, Stanisław Nagy, Barbara Gąciarz and Thomas Sattich

Routledge
Taylor & Francis Group
LONDON AND NEW YORK

First published 2026
by Routledge
4 Park Square, Milton Park, Abingdon, Oxon OX14 4RN

and by Routledge
605 Third Avenue, New York, NY 10158

Routledge is an imprint of the Taylor & Francis Group, an informa business

British Library Cataloguing-in-Publication Data
A catalogue record for this book is available from the British Library

ISBN: 9781003863014 (hbk)
ISBN: 9781003863045 (pbk)
ISBN: 9781003607151 (ebk)

DOI: 10.4324/9781003607151

Typeset in Times New Roman
by KnowledgeWorks Global Ltd.

Contents

Figures and Tables

Figures

Tables

About the Authors

Dariusz Wojakowski is a sociologist, professor in the Department of Society and Technology Studies, and Vice-Dean for Cooperation and Students at the Faculty of Humanities of the AGH University of Kraków. As of 2019, he is the Secretary of the Commission on Threats to Civilization at the Polish Academy of Arts and Sciences. His research interests include regional and borderland studies, sociology of local communities, social acceptance of energy transition and qualitative methodology. Recent publications are devoted to: borderland community resilience; territorial branding; borderland representations and social acceptance of CCS. These publications have appeared in, among others: "International Journal of Greenhouse Gas Control"; "Borderland. Social Studies"; "Konteksty"; "Polish Sociological Review"; "Sociologia. Slovak Sociological Review"; "Regional Science, Policy and Practice".

Oluf Langhelle is Professor in political science at University of Stavanger, Department of Media and Social Sciences, Norway. He is currently Head of the Department. He received his Dr. Polit. degree from the University of Oslo, Norway. His research has focused on the concept of sustainable development and follow-up, strategies for sustainable development, environmental politics and policy, including oil and gas policies in the Arctic, transitions toward low carbon societies, focusing on Carbon Capture and Storage (CCS), and the Doha Round Negotiations in the World Trade Organization (WTO). His current research interests focus on climate politics and policies, climate mitigation strategies, energy transitions, renewable energy, battery electric vehicles and transition theory.

Mateusz Stopa, PhD, sociologist, during AGaStor project worked as a postdoctoral fellow at the Department of Media and Social Sciences, Faculty of Social Sciences (University of Stavanger, Norway). He was responsible for qualitative research and analysis of secondary data (strategic documents, scientific and gray literature) and in-depth individual interviews with the representatives of regional institutions in the West Pomeranian

Province. His research interests include: sustainable development, social perception of energy transition, societal basis of innovativeness. He explores these topics from local and regional perspective. Currently, he is a methodologist and project manager at SOLIDATA (research company, Poland).

Stanisław Nagy is a professor at the Department of Gas Engineering, AGH University of Kraków, Poland. He is the Secretary of the Commission on Threats to Civilization at the Polish Academy of Arts and Sciences. He holds a PhD in Gas Engineering. His research focuses on thermodynamics, reservoir and gas engineering. It also deals with the energy transition process (in the context of reducing emissions through implementing CCS) and energy security. His interests also lie in engineering work related to low-temperature geothermal systems. He is the author and co-author of numerous monographs and articles.

Barbara Gąciarz is a professor at the AGH University of Krakow and has a PhD in sociology and economics. She held the position of the Dean of the Faculty of Humanities from 2015 to 2024. She was Head of the Department of Sociology of the Economy and Social Communication of WH AGH for several years. For many years she was associated with the Institute of Philosophy and Sociology of the Polish Academy of Science (IFIS PAN) in Warsaw. She specializes in economic sociology and the sociology of institutions. Her studies focus on institutionalization processes, local government functioning and the policy regarding the support for people with disabilities. She is the author and co-author of many books and articles, including "Local Government and Development. Institutions – Citizens – Empowerment", "Institutionalization of Territorial Self-government. Actors and Effects".

Thomas Sattich is Associate Professor at the Department of Media and Social Science, University of Stavanger, Norway. He holds a PhD in Political Science and Economics. The focus of his research is on the geopolitics of the energy transition, including the international implication of renewable energy and energy transmission grids, industrial policy and energy security. Sustainability education is another subject covered by his work, for example, through the development of Geovania, a simulation that allows students to explore the geopolitics of the energy transition.

Introduction

Idea of the Book

This book explores the reception and attitudes toward a new and novel technology that combines underground storage (UGS) for natural gas with CO_2 storage in Poland. Our case is the AGaStor project (AGaStor project – advanced gas and carbon dioxide storage in aquifer), a project that looks at possible new places for seasonal storage of natural gas. The defining technological feature of the AGaStor project is the utilization of CO_2 as a cushion for natural gas storage (with the purpose of increasing the efficiency of recovery of stored gas). The utilization of CO_2 makes this a novel Carbon Capture Utilization and Storage (CCUS) project, a technology that is not well known, not even among energy experts. The AGaStor project itself is more precisely described as a "utilization and storage project" (US) since the capture of CO_2 is not a direct part of the project but will have to come from somewhere else. As captured CO_2 is necessary though, in the following we refer to the project as *a potential CCUS project*. From a social science point of view, this means merging a more controversial technology, CCUS, with a more uncontroversial technology, underground natural gas storage (UGS).

The geographical focus of this study is on the West Pomeranian Province of Poland, a region where energy has a high importance, and where new infrastructures have been introduced over the past years, particularly LNG, natural gas storage and wind parks. The AGaStor project would add a new element to the ongoing changes in the regional energy infrastructure. Against this backdrop, the AGaStor project allows us to explore the social and political repercussions that come with the possible introduction of new combinations of CCS-related technologies. The research is therefore part of what Devine-Wright and Wiersma (2020, 1) calls upstream research, that is, "research that investigates the 'place-technology fit' of a potential … energy project before it is proposed". While Devine-Wright and Wiersma (2020) focus on a renewable energy project, our focus is on the use of CO_2 for underground storage of natural gas. The empirical part of the book is guided by the theoretical concept of "place-technology fit" as described by Devine-Wright and Wiersma (2020, 2), that is "symbolic beliefs about the local landscape … and the degree to which a given technology project is interpreted to be suited or 'out of place'

DOI: 10.4324/9781003607151-1

in that context". In other words, we are looking for meanings associated with both place and technology.

J.D. Peters (1997), however, writing about bifocality (i.e., *countable and the uncountable*), pointed out that social research, wherever it is carried out, should always be considered in both global and local contexts. In our view, the complex contemporaneity of the 21st century causes this spectrum of viewing social problems to "prismatize" into many intermediate dimensions – global and continental (the EU) at the macro scale, and regional (national) and local at the micro scale. The research devoted to the social acceptance of CCS and UGS technologies in Poland can therefore resonate on all those dimensions in which the problem of energy transition, or more generally economic and social development, appears. Of course, from each of these perspectives, they can be read differently, but basically as a transitional stage in the process which leads to zero net carbon emission. We are interested in technological solutions and recognizing the social situation at this particular point in time and space of the energy transition, when we are still heavily invested in fossil fuels.

From this perspective, studies of public acceptance of CCS and UGS in Poland are an exemplification of the overall context of energy change. It allows us to recognize how the public could potentially react to the implementation of any new technology related to the energy economy. The book shows what happens when engineers propose a new technology recognized by the scientific community as efficient and safe, politicians see it as a tool for energy-safety policy and the public is confronted with the need to make a statement about both the utility and value of something it is encountering for the first time. Turbulence at the international level of politics will probably require further and rapid and innovative adjustments of the existing energy system, particularly if the ongoing transition to a sustainable energy sector is to be maintained. Understanding public acceptance of novel solutions is, therefore, crucial to ensure energy security. In particular, this includes solutions that enable progress toward sustainability and safe supply with natural gas.

The energy economy, at all levels of the social prism, faces challenges that constitute a related and inseparable triad: transition-safety-accessibility (Arias et al., 2023). Although it is sometimes understood in different ways (Child et al., 2018; Pasman et al., 2023), it should be the main reference for recognizing the importance of social acceptance of new technologies in energy.

The overall aim of this book is therefore to understand the possibility of community acceptance of this novel and hardly known combination of energy technologies. We are thus conducting this study in anticipation of the future use of the technologies mentioned. Devine-Wright and Wiersma (2020) find that relatively few studies on community acceptance have taken place prior to specific projects being proposed or implemented. They argue, however, that such studies could be highly beneficial, as a priori studies "may have greater impact on policy makers, industry and civil society since findings can inform pathways of local low carbon energy deployment" (Devine-Wright and Wiersma, 2020, 2). In the case of AGaStor, it could be argued that community acceptance would be vital for the project (as well as similar ones) to materialize.

In the following, we will explore the reception of the possible implementation of this technology with both quantitative and qualitative data from two of the main potential locations, Marianowo and Choszczno (Suliszewo) in Poland. Our main research questions are first, to identify the prior awareness and knowledge of CCUS. Second, to identify and discuss possible receptions of this novel technology, and third, to identify the factors that might influence and explain individuals' attitudes and possible acceptance of new technologies. In this part, the economic, social and cultural conditions are taken into account as the possible factors of those attitudes. What we find is that recent emotions and past experiences with development projects of different kinds influence the attitudes toward new proposed projects in a substantive way.

Methodology

Our research is based on data gathered from both qualitative and quantitative procedures. To analyze public discourse, we also use desk research which lets us gather some existing data.

As for the analysis of Polish public discourse on CCS and CCUS, three different procedures had been implemented: (a) content analysis for official governmental strategic documents, (b) Google Search in anonymous mode for the content of traditional media and (c) Google Trends analysis for social media content along with search on Facebook. The most complex had been content analysis of the documents – seven documents ranging from 104 pages up to 418 pages (in total – 1,559 pages of content). There were five main steps of the quantitative content analysis of governmental perspective in public discourse on green transition in Poland and perception of CCS as a possible element of the strategy:

1. All documents have been searched with ATLAS.ti analytical software to identify all words that occur at least once in at least one of the documents. This is the only way to cope with different approaches of authors and possible different terminology used for the same issues. Thanks to this, the list of 37,325 words has been generated;
2. The generated list of words has been carefully read to group any words (in different grammatical variation) that may refer to the topic of the analysis. This action has led to the compilation of a list of 250 words (taking into account grammar declination and different variations of the words);

3. All these 250 words have been precategorized into codes – abstract and general equivalents, as fully as possible reflecting the character of the component words. These codes are supposed to refer to the main problem area of the research;
4. However, it is obvious that one word may have different meanings in different contexts. Therefore, all codes have been checked (searched) in all the documents whether they refer to fragments of the text that in any way relate to the main problem area of the study. For example, word "emission" may refer to greenhouse gases emission or to emission of the shares on the stock or emission of the voice – while the first meaning relate to the topic of the research, two others not necessarily;
5. All occurrences of 23 codes in the texts of all strategic documents have been counted. Of course, only those that in any way had related to the topic of the research. If English translations of two or more different Polish codes have been exactly the same, then these codes have been merged and counts summed up. During this action also, power citations have been gathered – meaning such fragments of the content that most fully illustrated the meaning of a given code as assigned by the authors of a given text.

The content of the archives of traditional media had been searched in anonymous mode to avoid any possible bias of the results due to the previous search engine (Google) usage in a similar topic area. The search was conducted for the Polish language in the time period from 01.01.2020 to 31.12.2023. The keywords selected for the search were "carbon capture" and "carbon storage". Every result associated with any traditional media: newspaper, radio station, TV station, was checked.

The same search terms had been used in Google Trends analysis in order to verify to what extent the topic of CCS/CCUS actually generates any spontaneous interest. The application of this procedure is based on two assumptions (confirmed by independent reports on the level of computerization and digitization of Polish society): (a) the vast majority of society has access to the Internet, and (b) this majority uses a unified tool to search for any information online. Consequently, if the entry is searched for, it is an indirect indicator of interest in a given issue. Of course, the result for only one issue can be misleading. Therefore, a point of reference was used in the form of an issue that has been strongly discussed in Poland for a long time on various levels, and which is related to the green transformation and also appears in the analyzed strategic documents – i.e., photovoltaics. Regardless of the results of the Google Trends analysis, Polish Facebook resources were additionally checked for possible discussion groups devoted to the topic of CCS/CCUS. Such groups exist in large numbers and are active, for example, in English or German – constituting a platform for the exchange of information and opinions on this topic. The possible existence of similar Polish discussion groups could provide further content for analysis.

The first quantitative research was a regional representative survey. The sample used in the survey is 695 adult respondents from West Pomeranian Province. CAWI (computer-assisted web interview) was made by survey company Ariadna on 15–22 September 2021. The questionnaire contains 10 main questions and 12 independent variables, which allows distinguishing 87 items (indicators) for statistical analysis. Subjects of the questions are divided for: (a) knowledge about UGS and CCUS – 3 one-item's questions; (b) CCUS acceptance: 2 questions – 14 items; (c) risks and benefits connected with new technologies: 2 questions – 12 items; (d) NIMBY factor in CCUS context: 1 questions – 6 items; (e) system of values (pro-ecological worldview): 2 questions – 18 items; (f) independent variables: demographic, economic, connected with the worldview: political orientation, religiousness, civic activism. All quantitative data were analyzed with the SPSS statistical program.

The qualitative research was based on deliberative focus group interviews (Rothwell et al., 2016). Generally, group interview is defined as the technique where the researcher gathers all necessary information by and from cooperation and interactions in a group of participants (Lisek-Michalska, 2013, 16). According to Lisek-Michalska, interactions within the discussing group and active role of the researcher constitute the context of data collection. The technique of deliberative group interviews is an "effective modality for obtaining in-depth data within these newer and more complex [...] research contexts, [because – auth.] more information and education is required of the participants to engage in a productive focus group discussion". Two key additions to the typical focus group experience help facilitate a deliberative discussion focus group approach and include: an educational presentation prior to the group discussion and the addition of an expert co-moderator (Rothwell et al., 2016, 3).

The scenario of the discussion consisted of four parts:

1 Introduction – information on the project, participants representing the research unit, discussion about interviewed participants and their local communities. There was also a short and general discussion on innovativeness as a potential characteristic of the local community.
2 CCUS and UGS technologies presentation – presented by the experts in the field of geology and engineering sciences.
3 Discussion on these technologies with the presenting experts – moderated by the sociologist.
4 Discussion on the technologies and their use in local communities – conducted and moderated by the sociologist but without technical experts' presence.

The AGaStor project explores the localization of CO_2 facilities in the West Pomeranian Province either in Marianowo or Choszczno commune (two bordering communes located in one geological system). There were three meetings conducted within this scenario (as classic single sessions; 6, p. 18) in May and June 2022. Two of them were in the potential localizations of a CO_2 storage (Marianowo and Choszczno), the third was in the capital of the province (Szczecin). There were 18 representatives who participated in those deliberative group interviews.

The main assumption on the character of the participants was that they should represent categories of local and regional stakeholders: local authorities and specialists in environmental protection issues (officials and local government representatives, signed Rp), journalists, members of NGOs and local institutions (all signed Rs). Such a selection was proper for the technique and the groups can be treated as affinity groups. "One selects both groups – by defining their specificity, as well as participants – fitting planned selection criteria. Besides all socio-demographic conditions, the common experiences are crucial in selection, meaning that one must be confident that each of the participants meets a certain type of substantive criterion" (Lisek-Michalska, 2013, 22). During the meeting, there were up to four recording devices and all soundtracks are backed up and stored on hard drives and in the cloud. Each discussion lasted approximately 120 minutes.

The crucial moment of the deliberative focus group interview (DFGI) technique is the presentation of the CCS and UGS technologies. This presentation was made by a professional researcher in the technical sciences based on some key information. First element of that presentation was the information about the realization of CCS projects in the world. Next information was about the specificity of the CO_2 underground storage. The information for participants also contains some details about: the geological parameters of a gas storage in aquifer; CO_2 trapping mechanisms and functioning of CO_2 storage facility in saline aquifers. The presentation stressed that the guiding concept of the AGaStor project is based on the synergy between underground storage of natural gas storage and geological storage of CO_2. Combining CCS technology and UGS operation may reduce CO_2 emission into the atmosphere, as well as reduce the cost of UGS cushion gas. The audience was explained that, in addition to suitable geological conditions, it was a favorable location with respect to existing gas pipelines that determined the choice of the given structure. During the presentation, the participants could ask about details, and after it was finished, they discussed those topics with the scientists who presented the information. Finally, the participants formulate the opinion about the technology after that discussion, without the presence of the experts.

The qualitative analysis of the data was made in the MAXQDA 2022 program. The data was coded in the way proposed by K. Charmaz (2003) and had three stages: (1) incident to incident initial coding; (2) focused coding; (3) preparing theoretical memo-writing.

The evaluation research took place on 27 April 2023 and 19 May 2023. Meetings were held with residents of Suliszewo and Marianowo, which used the communication strategies developed in the qualitative research. Those meetings have a deliberative character as participatory action research (PAR) which could be treated "as a platform to overcome the domination of knowledge associated with conventional research methodologies. PAR allows for a more democratic process of knowledge development through the voices and experiences of local people" (Nielsen et al., 2022, 15). After presentation and discussion about the project and the proposed technology, participants were asked to fill out an evaluation questionnaire. This method yielded 31 questionnaires.

Structure of the book

The following first three chapters have a distinctly theoretical profile. Their purpose is to show the various contexts of the research. Chapter 1 is a review of the literature on the social acceptance of new technologies with a particular focus on CCUS. We look to the current theoretical findings as a source of inspiration for interpreting the research results that we have obtained. Thus, we reach here for sociological and political studies. Chapter 2 is intended to familiarize the reader with the geological and physical aspects of the technological solutions adopted in the AGaStor project. This chapter also explains to what extent CCUS is a technology recognized by the European Union as a way of mitigating CO_2 emissions in the process of the full energy transition of our continent. Chapter 3 is an analysis of energy-safety policy from a social and political perspective which indicates the importance of the energy transition for the security of the EU's Northeastern region.

The two following chapters analyze the national and regional contexts. Chapter 4 describes, based on an analysis of foundational data, the public discourse on CCUS in Poland. It identifies the actors in the discourse, the main media channels and contexts related to other renewable energy technologies (RET). Chapter 5 characterizes the region and local communities that could become potential locations for UGS presenting economic, political and demographic characteristics of the West Pomeranian region.

The last three chapters report on the results of sociological research conducted in the AGaStor project. Chapter 6 reveals the demographic and worldview contexts of acceptance and rejection of CCUS based on quantitative research. The research shows to what extent technological ignorance builds a sense of uncertainty. Ignorance and uncertainty about new technologies are universal phenomena in the modern world. The chapter focuses on what factors reduce feelings of uncertainty, immunize people against ignorance and enable them to make decisions. The qualitative research in Chapter 7 shows that distrust, rejection and trust can stem from past experiences. The filter through which new technology is evaluated is local needs and problems. If local needs diverge from what is offered, then we are dealing with indifference. The chapter shows the various motivations for rejecting (or accepting) new technologies. Chapter 8 focuses on the negotiability of social attitudes and examines how effective CCUS communication tools are. It identifies trust-building mechanisms and, based on evaluation surveys, indicates areas of their effective impact.

AGaStor advanced gas and carbon dioxide storage in aquifer
CCS carbon capture and storage
CCUS carbon capture utilization and storage
EOR enhanced oil recovery
DFGI deliberative focus group interview
NIMBY Not In My Backyard
OWMA only with my acceptance
RET renewable energy technologies
UGS underground gas storage
WIMBY Why In My Backyard

Note that in the book we use terms "CCUS" or "CCS and UGS" convertible if we appeal to the technology used in AGaStor project.

1 What Do We Think about Social Acceptance of New Energy Technologies?

Global risk society: the new place of technology and science in public discourse

The initial theoretical approach is the concept of a global risk society by U. Beck. This concept seems particularly valuable in analyses devoted to the social acceptance of new energy technology. It deals with the role of science and technology in the modern world, but also shows the mechanisms of social reactions to modernization. Problems of the "risk culture" and "constructed uncertainty" are crucial to understand the social reaction to new technologies (Beck, 2012, 187–8). In studies devoted to society and technology, those issues relate to the category of uncertainty (Beck, 2012; Nicolas et al., 2021; Nowotny et al., 2001). Uncertainty always functions in a categorial framework together with risk estimation, ignorance and decisions making. These categorial constructs could be used for analysis of scientific discourses. Here the best example is the question on climate uncertainty (Nicolas et al., 2021) or on the implementation of new technologies (Otto, Sprenkeling et al., 2022).

The most interesting empirical data for these theoretical studies could be found in the area of technological ignorance, local rejection and social attitudes. First one is connected to Beck's idea of a reflexive ignorance. Otto and Gross (2021) note that the communication of such ignorance, which heavily distorts the reception of technology, depends on communication styles. In contrast, other studies show that new social technologies are recognized as un-resolved or long-term uncertainties (so-called unknown-unknowns; Pidgeon et al., 2012, 4181). Pidgeon also noted that the opinion/decision-making process involves other competencies of the respondents involving values and trust:

> The study of public perceptions and beliefs can provide insights into the value and ethical concerns not ordinarily considered by scientists and engineers, or factors such as trust in risk management, all of which are important for the governance and control of risks, but which often prove

DOI: 10.4324/9781003607151-2

> more difficult to express in formal risk–benefit assessments. Perceptions also matter because *beliefs (even when they turn out to be completely wrong) can have real consequences in the world.*
>
> (Pidgeon et al., 2012, 4177)

Deciding what is right or wrong (or completely wrong for that matter), however, is not straightforward and can be a difficult question to answer. Risk society is not only about the distribution of "goods" (wealth) or "bads" (pollution, contamination and other by-products of production). Risk society also comes with "a loss of faith in experts and science to predict and protect people from these technological hazards", and "increasingly competing knowledge claims (erosion of expert consensus) regarding the management of technological environmental hazards" (Baxter, 2020, 305). Still, as Baxter points out, experts and expert publications "are actually called upon by all stakeholder groups for and against hazardous facilities/technologies and the combined effect of the way messages are deployed can either amplify or attenuate concerns about environmental health threats" (Baxter, 2020, 305).

The erosion of expert consensus makes knowledge claims part of the contestation in the societal and political landscape, where societal groups and communities may focus their concern on some hazards and not others, and where "those choices may not align well with the way 'experts' assess those same hazards" (Baxter, 2020, 305). Moreover, it impacts on community acceptance by which we mean "the specific acceptance of siting decisions and renewable energy projects by local stakeholders, particularly residents and local authorities". Upham et al. (2015) define acceptance as "a favourable or positive response (including attitude, intention, behaviour and – where appropriate – use) relating to a proposed or in situ technology or socio-technical system, by members of a given social unit (country or region, community or town and household, organization)". However, in the literature, the concept of technology acceptance is defined in various ways, which indicates its multidimensionality. The definitions and content of this category are the subject of the next section of the chapter.

Issues of social acceptance in CCUS

Among contemporary studies, Gordon et al. (2022) take the most clear-cut stance on the acceptance of new energy technologies. They try to clearly separate social acceptance and public acceptance:

> social acceptance should therefore be viewed as "a socially constructed outcome" – as opposed to a "one-way process" of adoption or rejection of technology innovation – in which interests become "mutually aligned" and complement existing socio-cultural structures. Furthermore, the object of social acceptance research should be "new phenomena, new ideas,

> policies, and projects". This reconfiguration of the key components of social acceptance, marking an explicit distinction from public acceptance, exemplifies the self-reflective nature of third wave scholarship and the drive towards more integrated research approaches.
>
> (Gordon et al., 2022, 12)

From such a perspective "public acceptance is only part of multidimensional and multi-actors relations which also contains public acceptance as the sum of individual behaviour, attitudes and tolerance" (Gordon et al., 2022, 13).

Another proposition could be founded by Dütschke et al. (2019). This idea is based on differentiation of socio-political, local and market acceptance as facets of acceptance (Wüstenhagen et al., 2007):

- According to those authors "socio-political acceptance refers to the general societal climate towards a technology or innovation within a society, i.e. it relates to typical discussions about a topic or social desirable opinion. [...] Socio-political acceptance is shaped and mirrored by opinion leaders, poll data, media and alike" (Dütschke et al., 2019, 9).
- "*Community acceptance* is a concept that is most relevant for siting decisions and refers to the attitudes and behaviours exhibited by neighbours of installations or others somehow affected by an innovation or technology without actually using it. In case of CCUS this includes the full chain, i.e. the CO_2-source, the transport by pipelines, trucks or ships, etc., and the installations for CO_2 usage or storage. [...] most discussions around social acceptance of CCUS have come up around storage sites" (Dütschke et al., 2019, 9). As Wüstenhagen et al. (2007, 2685) point out, community acceptance "refers to the specific acceptance of siting decisions and renewable energy projects by local stakeholders, particularly residents and local authorities".
- "*Market acceptance* finally refers to the acceptance of a technology that is manifested by market actors, i.e. supply and demand side as well as intermediate actors like installers, consultants etc." (Dütschke et al., 2019, 9).

What is important, both cited studies present the perspective as distinguishing between different levels of acceptance analysis: the national level (defined as social acceptance), the local level (defined as community acceptance) and household/organization level (defined as market acceptance).

All these propositions are rooted from research experience and institutional needs of entities interested in investment in storage facilities. They have an important impact on thinking about acceptance of new energy technologies, and all are useful in planning research. But in our opinion they could produce some misunderstandings because they omit basic sociological categories and theories that describe the fundamental nature of the social world – maybe because they are too categorically focused only on Actor Network Theory, which is rather a methodology of analysis not a theory as such. The only obvious and strong sociological category which appears in these definitions is "attitude". It is worth remembering that the components of every developed attitude are: meaning/image of the object, feelings connected with it and tendency to act (Blumer, 1936, 515–6). Therefore, the attitude is always social construction/definition of an object (Blumer, 2007). An individual as well as a group could build such constructions toward every social and natural subject (things, people, groups, spaces, ideas, technologies, etc.), not only "new phenomena or ideas". This assumption fits partially to Gordon and his co-authors' idea of social acceptance as "socially constructed outcomes" of the interests of many social actors (rather than structures). But such definition also fits the idea of public sphere which – according to J. Habermas – is "a domain of our social life where such a thing as public opinion can be formed, [where] citizens [...] deal with matters of general interests without being a subject of coercion [... to] express an publicize their views" (Habermas, 1991, 398).

Therefore, Gordon's definition describes a more public than social dimension of acceptance despite it emphasizes very accurately the processual and constructive nature of acceptance. Probably this problem is connected with the fact that the components which Gordon and his co-authors recognized as social acceptance (multiple actors and scales; dynamic processes and coevolving actors relationships; Gordon et al., 2022, 13) are just natural sources of "socially constructed outcomes" – the acceptance of new technology. In other words, they describe the way how such acceptance is created and how it changes in time. But these features do not indicate the content of the acceptance and the direction of its changes. Sociological research needs the indicators of social perceptions, opinions and imaginations about new technologies. Such indicators could be found in Gordon's public acceptance definition and Wüstenhagen's and Dütschke's definitions of community acceptance (behaviors and attitudes).

Dütschke's definition of socio-political acceptance is just in the center of the "public" idea because it points to the main phenomena which "are present" in the public sphere and could be "expressed and publicized" (public opinion, experts, media). These phenomena could be researched and analyzed by the methodologies that appeal to public discourse as the central analytical category (see Chapter 4).

Searching for the "content" of social acceptance of carbon capture, utilization and storage (CCUS), it is worth noting the analysis of a broad literature review conducted by Tsvetkov and colleagues (2019, 3). They identified the four most popular themes (knowledge; acceptance of CCS and preference between technologies; governmental policy and interaction between stakeholders, benefits and risks perception) and four frequently occurring themes (socio-demographic factors, trust, awareness, NIMBY).

These issues are very different and, as the authors themselves note, not necessarily disconnected. Indeed, the issues of knowledge and awareness can be treated together. Their high position in the literature is due to: "The largest number of articles devoted to the influence of knowledge about the nature of CCS on the public perception, which is fair for almost all countries, because it is still little-known technology" (Tcvetkov et al., 2019, 3). This is why we treat the knowledge and awareness issues as strongly tied with the risk society theory described in the previous section of the chapter. We highlighted the research of Pigeon et al. which, in the most general terms, addresses what importance values and ignorance have in responding to new technologies.

"Governmental policy and interaction between stakeholders" belongs rather to the separate sphere of public acceptance, as we have delineated above and described elsewhere in the book (see Chapter 4). This is the content that cannot be analyzed by opinion polls alone, but should be the subject of triangulated sociological research (see Denzin, 1979).

In general, the "acceptance of CCS" is a category inseparable from such categories as "trust", "benefits and risk perception" and "NIMBY". On the other hand, each of these three categories is well-established as separately studied components/content of social acceptance of new technologies. Trust is that aspect which can be primarily associated with the way new technologies are communicated. This issue is addressed in the last part of this theoretical chapter.

The NIMBY phenomenon, on the other hand, is revealed in research as a very complex phenomenon. Originally, NIMBY encompassed the range of behaviors and attitudes that reject new technologies in opposition to Upham's "favorable or positive response". Now, in the context of understanding social acceptance as a dynamic and multidirectional process, it falls within the spectrum of such attitudes and activities which creates that process.

Social acceptance of new technologies as spectrum of social attitudes and activities

Thinking in terms of the social acceptance of technology – which presupposes both the participation of individuals and communities in the decision-making process and the possibility of rejecting such participation – implies an assumption of the democratic nature of processes related to energy transition and energy-safety policy: "[in the process of the energy system transformation] democratisation has increased in importance and will become more important in the future, especially in the context of acceptance issues" (Wagner and Götz, 2021, 3).

As Wagner and Götz write:

> The democratisation of the energy industry covers a very broad field. This includes the possibility of many individuals becoming electricity producers themselves; however, this also requires addressing the desire for social

> participation in energy issues. [...] It is also an expression of growing distrust of large energy supply companies and a system for the provision of services of general interest that is predominantly based on shareholder value.
>
> (2021, 11)

Democratization is especially associated with RET and decarbonization (Clulow and Reiner, 2022; Haas et al. 2022). Gordon et al. write that:

> Whereas some technologies such as nuclear power have traditionally been associated with political elitism and technocratic ideologies, renewable-based energy systems may present opportunities to create a more democratic, decentralized and community-based energy future. This prospect has seen "clean energy" politics increasingly associated with the notion of "energy democracy", demonstrating the far-reaching implications of energy choices on the socio-political landscape.
>
> (2022, 9)

Gordon and colleagues note also that "critical approaches depart from previous scholarship by unpacking how RET deployment and the decision processes behind it are often apolitical, unjust and undemocratic, or just another materialization of the neoliberal capitalist system" (2022, 9). Despite such perspective on energy transition processes, we must agree that democracy makes this transition more complicated because "decision-making processes often involve negotiation, compromise, and consensus-building, which can slow down or weaken the implementation of ambitious climate policies. As a result, indicators representing individual and civil liberties generally rank lower in terms of their effectiveness in mitigating greenhouse gas emissions" (Tsur, 2025, 2).

It leads us to explore and discuss social acceptance as a broad spectrum of attitudes toward new energy technologies. There are several possible and legitimate reasons not to accept, support or tolerate new energy infrastructure in specific places. Acceptance and support are also different. Support implies a "more active and favorable position", while acceptance is more related with passive reception (Batel et al., 2013, 4).

Batel (2020) divides research on the social acceptance of RET and associated infrastructures in three successive waves, *normative*, *criticism* and *critical*. Although our project, strictly speaking, is not an RET project, it is still about social acceptance of a potential investment that aims to contribute to a green transition. The waves identified by Batel are therefore highly relevant. The first wave is loosely attached to the debate about NIMBY (*Not In My Backyard*), where the physical proximity explains opposition or not. Wüstenhagen et al. (2007, 2685) describe the debate surrounding NIMBYism "where some argue that the difference between general acceptance and then resistance to specific projects can be explained by the fact that people support renewable energy as long as it is not in their own backyard, while others argue that this is at least an over-simplification of people's actual motives" (see also the critique of NIMBY by Devine-Wright, 2009). Batel describes this first wave as normative as the main line of inquiry is to characterize opposers and supporters, with the goal to overcome opposition (Batel, 2020).

The second wave, *criticism*, comes as a reaction to the first, and raises the need for an alternative understandings of opposition. Batel (2020, 3) describes the main line of inquiry in this wave to examine the socio-psychological and community factors that affect opposition, including perspectives on social justice. The goal or societal implications of the studies focus more on the understanding of opposition and how it would be possible to ease the transition (by increasing acceptance). The third wave, *critical*, is more diverse, challenging and criticizing ideological stances, focusing on power relations and shifting the focus away from research on acceptance to research on *other* responses, such as support, tolerance, indifference; questioning the need to build and why it should be built at the specific location; the relation between expert-political and lay systems and how democratic those relations are. In short, the critical wave asks the question if opposition *should* be reduced or overcome or not (Batel, 2020).

Thus, there are many possible responses other than acceptance, and there are also several possible and legitimate reasons not to accept, support or tolerate new energy infrastructure in specific places.

This diversity of responses is described by Gordon et al. (2022). They highlight three components of public acceptance: behavior, attitudes and tolerance. Based on this, they create a model of diversity of responses – including both acceptance and rejection of technologies related to energy transitions (Figure 1.1).

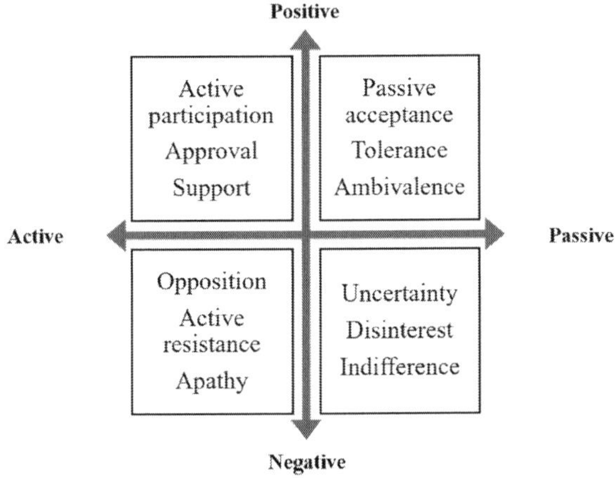

Figure 1.1 Gordon et al.'s social acceptance matrix [2022, 13]

Authors focused on the liquidity of these attitudes: "Consumer responses can fluctuate freely between positive–negative and active–passive ends of this spectrum, corresponding to the adoption or rejection of new behaviours and practices" (2022, 13). The bottom of this model indicates a variety of motivations for rejecting new technologies. While the active-negative quarter of the matrix describes typical NIMBY attitudes (with opposition and active resistance), also in the negative-passive quarter the attitudes that ultimately reject investment in new technologies could be found. It corresponds with the WIMBY (Why In My Backyard?) concept developed by Svartdal and Kristoffersen which also focus on *another* response, namely indifference. WIMBY is defined as "local indifference to take part in changes in local energy landscapes at the community level", and it is seen as a problem across sites: "Even if local publics are in favor of green transitions and recognize the challenges of climate change, they don't necessarily see the relevance to their community. This has several implications besides potential foot-dragging, as pilot and demonstration projects struggle to anchor socio-technical measures in communities". In essence, "energy transitions are being delayed not by active resistance, but by a passive lack of engagement" (Svartdal and Kristoffersen, 2023, 1).

Communicating new technologies: strategies and styles

The last theoretical topic popular in the social acceptance of CCUS research is the strategy of communicating the technology to the public (Otto and Gross, 2021). Most of the studies concentrate on one-way communication in traditional media (mainly the press) and analyze the information prepared and communicated by experts and scientists to public opinion. The analyzed press articles also contain public reactions to these information and plans in the form of statements by formal and informal representatives of local communities (during planned or already ongoing local social protests). Of course, the category of experts or scientists is not homogenous: there are both supporters and opponents of the technology, representing industry, scientific institutions and NGOs.

The review of the literature reveals differences in communicating on CCUS, depending on the category of stakeholders that had been researched (Table 1.1).

For declared opponents of CCUS technology, the hierarchy of available solutions to the climate change issue is as follows: on one end of the scale – renewable energy sources, then sustainable energy, CCUS technology and conventional sources, nuclear energy – on the other end of the scale. In such an approach, CCUS as a technology is regarded as being only a fig leaf, it is perceived as "greenwashing" of the main cause of climate change – conventional sources/fossil fuels, and at the same time it is perceived to consume

Table 1.1 Framings assigned to specific discourse parties

Stakeholders' category	*Framing of CCUS*
Industry representatives	"Faith in innovation" "Value instrumentalization" "Status quo maintenance" "Omissions"
Experts and scientists	"Technological challenges" "Administrative and legal aspects" "Changes and risks" "Societal relevance and need for action" "Communication"
Public opinion	"CCUS as a trigger for global considerations on lifestyles and decision-making process" "Humanity's relationship with nature and environment" "CCUS as an element of value chain"

Source: Own literature review.

valuable financial and knowledge resources that could be devoted to a more intensive development of renewable energy.

It is worth stressing that usually CCUS projects are located in less developed, less populated areas, with smaller, more rural local communities facing serious socio-economic challenges. Definitely, this is the case presented in this book. In fact, low income and low socio-economic status and the resulting expectation of compensation (possible benefits to the community) but also past experiences with government are far more important than possible concerns about the risks of technology itself (Bradbury et al., 2009). This particular study concludes that

> (…) key management questions for the public are: How can we have a say in what happens? Who is in charge? Will the process be fair and will anyone listen to us? What will happen if something goes wrong? Can we trust the project developers and the government to take care of any problems – what have our previous relationships with these entities shown us? What is the benefit to our community? How does the proposed project fit into or improve our way of life?
>
> (Bradbury, 2009, 4671)

Therefore, from the local community's point of view, even the "Not In My Backyard" (NIMBY) strategy is rather a mechanism of emancipation and articulating community's subjectivity, and a strategy for articulating fears of distrust and inadequacy of empowerment. NIMBY strategy can be generalized as "the protectionist attitudes of and oppositional tactics adopted by community groups facing an unwelcome development in their neighbourhood" (Dear, 1992, 288). Such definition does not exclude potential positive attitude toward proposed development – the resistance is due to the mentioned inadequacy of empowerment, and the decision considers the only manageable space, its own backyard.

In other words, the emancipation of local community should be based on equal and active position of its representatives in the process of building an informed consent about what is known and what is unknown, so local stakeholder groups may decide to start to act in spite of enormous non-knowledge (Gross, 2007, 752). The local community must agree on what is unknown and take it into consideration while planning the project (Gross, 2007, 753). The trust is built more on the guarantee of shared responsibility for unforeseen consequences (as the strategy for coping with uncertainty) than on delivering excess knowledge.

Another literature review of papers on communicating CCUS (Otto and Gross, 2021) reveals four main areas of recommendations formulated by the researchers of this issue (Table 1.2).

These recommendations are treated as an evaluation model for our results. They do not constitute categories of analysis because it would be a type of pre-conceptualization, improper due to the inductive research techniques used in our study. Therefore, relating our results to these recommendations will be part of our conclusions.

Table 1.2 Categories of recommendations for communicating on CCUS based on literature review of 115 papers (2021)

Preparation	*Communication approach*	*Content and communication style*	*Support structures*
context-specific tailoring of the content right setting and channels for communication assembling competent communication team of independent local experts	establishing "positive" attitudes towards CCUS providing knowledge to an unaware public gathering knowledge on the community and relevant stakeholders communication opportunities and challenges transparent and egalitarian communication addressing justice, compensation and ownership avoiding "over-consulting" the community	benefits and risks, the latter ones carefully communicated comprehensible language with mapping of the message storytelling with links to everyday life experiences combining technical information with norms, values and beliefs use of visualizations	decision makers, politicians local media landowners or teachers high profile public figures

Source: Otto and Gross (2021).

Conclusions

This chapter provides a description and definitions of the categories that capture the phenomenon of social acceptance in a broad sense. We refer here to both these phenomena that manifest themselves in the public sphere (public acceptance) and those that are a collective and/or individual response to new technologies (social acceptance). Therefore, social acceptance of CCUS encompasses knowledge (and the ignorance), attitudes, behaviors, values, the tendency to perceive benefits and risks, and is ultimately manifested in trust, or lack of trust, in CCUS. At the same time, we assume that phenomena in the public sphere are not in opposition to social phenomena, but constitute some complex dynamic system of relations between different actors and social ideas.

It is worth taking into account that AGaStor is a research project in which two local communities have been confronted with the problem of forming their own social attitudes toward a potential investment in CCUS, which is geologically, technologically and politically possible. Although the project itself is not yet an investment, such a realization may be imposed on them due to factors external to these communities: geological (favorable formation of geological layers, making it likely that an underground storage facility will be leak-proof), strategic (proximity to the gas port in Szczecin and the key role of natural gas for Poland's energy system in the coming decades) and logistical (proximity to existing pipelines).

In such circumstances, theoretical frameworks presented in this chapter allow us to ask key research questions:

- How is the Polish public discourse around CCUS shaping up?
- What is its potential impact on public acceptance of CCUS at the regional and local levels?
- How is knowledge and ignorance about CCUS shaped in northwestern Poland and local communities where UGS investments are possible?
- What are the socio-demographic and worldview determinants of knowledge and ignorance about CCUS?
- What are the regional and local determinants of general acceptance of CCUS?
- How are attitudes toward CCUS shaped in the region and local communities surveyed?
- What might be ways to effectively communicate CCUS technology and build trust toward it?
- To what extent does communicating CCUS technology in participatory action research change knowledge about the technology?
- To what extent does communicating CCUS technology in participatory action research change attitudes about this technology?

Searching for the answers to these questions are presented in the empirical chapters of the book.

2 CCUS Considering Global Climate Policy and the European Green Deal

Introduction

The issue of the global environment first appeared in an institutional dimension on the international level in the form of the United Nations Conference on the Human Environment in Stockholm in 1972 (Brown Weiss, 2011). This conference established a basic framework for the issue for the next 20 years, introducing the official term "environmental policy" for the first time. In practice, the Stockholm Declaration "placed environmental issues at the forefront of international concerns and marked the start of a dialogue between industrialized and developing countries on the link between economic growth, the pollution (…) and the well-being of people around the world" (UN, 1972). In 1992 the United Nations Framework Convention on Climate Change was adopted (Brown Weiss, 2011). The Protocol to this Convention (Kyoto, 1997) explicitly assumed the reduction of greenhouse gas emissions by at least 5% (compared to 1990 levels) by year 2012 and introduced three mechanisms: emissions trading (ET); joint implementation (JI); and the Clean Development Mechanism (CDM). The appearance of the costs of emitting CO_2 meant the growing focus on CO_2 as an industrial waste and main greenhouse gas. The cumulation of this extremely politically complex project was Paris Agreement in 2015 (signed by 194 states and the European Union) that included commitments from all signatories to reduce their emissions "to achieve a balance between anthropogenic emissions by sources and removals by sinks of greenhouse gases in the second half of this century" (IPCC, 2022, 242; UN, 2015, 4).

In 2019, due to the significant probability of not achieving the assumed in Paris Agreement climate goals, the growing ecological concerns of EU citizens, but also the desire or even necessity to build a global economic advantage of member states in innovative, cost-intensive industries, including green technologies, the European Commission proposed the Green Deal (Charveriat, 2023). This strategy set the main goals for 2050: climate neutrality, circular economy, clean industry, healthier environment, more sustainable farming and climate justice and fairness. Among other actions there was the Fit for

DOI: 10.4324/9781003607151-3

55 package (2021) – a set of proposals to revise and update EU legislation to ensure the reduction of EU's net greenhouse gas emissions by at least 55% by 2030 (European Council, 2021).

However, given the current development of technology and the scale and dynamics of the global economy, the current initiatives to move from fossil fuel to renewables are not a sufficient solution to the issue of emissions (Quang, Milani and Zahra, 2023). "The EU's Industrial Carbon Management Strategy recognizes that without CCS, the EU will not be able to achieve its climate ambitions" (Jones, 2024). This has been a main factor of growing interest in CCS technologies as an additional and transitional solution for decarbonization. The technology of the separation of CO_2 from other exhaust gases had been known since the 1920s and had been introduced on an industrial scale in the oil industry since early 1970s – as a type of so-called Enhanced Oil Recovery (EOR), i.e., the use of CO_2 to increase extraction efficiency of crude oil from deposits (Thomas, 2008).

In 2024, the European Union launched another regulation (NZIA – Net Zero Industry Act) to support decarbonization processes – including CCS, requiring member states to geologically sequester a minimum CO_2 stream – for starters, designating it as mandatory for fossil fuel extraction companies (Table 2.1).

Europe wants to become the world's first climate-neutral continent by 2050. CCS/CCUS is becoming an important part of energy policy for three reasons:

- As a way to decarbonize the chemical, cement and steel industries;
- As the launch of a clean hydrogen economy (blue hydrogen is produced from natural gas with CCS);
- Achieving zero net emissions.

Table 2.1 Areas and sectors of the economy vs elimination of CO_2 emissions

Area 1	*Area 2*	*Area 3*	*Area 4*
Economic sectors which have technologies and can benefit from low-cost RES	Economic sectors that have low-emission technologies, but expensive	Economic sectors which currently do not have technology using RES	Economic sectors that never will be able to use of renewables
(29%)	(31%)	(19%)	(21%)
Energy	Maritime transport Rail transport (load) Private car transport Cargo transportation Heating and climate	Air transport Cement industry Steel industry Chemical industry	Fermentation of wastewater and waste Agriculture Landfills Wastewater treatment Deforested areas

The possibility of utilizing or recycling stored CO_2 (CCUS) is also linked to the closed-loop economy. CCU in industry refers to processes that capture CO_2 and then convert it into a new product. Such new products can replace or modify existing industrial materials, for example, carbon fibers and new cement. In this context, there is a need for CO_2 that can be captured in the process, stored and then used in such an economy.

CCS and CCUS technology

CCS was once considered the primary method for decarbonizing the power and energy-intensive sectors. Today, its potential is directed toward decarbonizing the economy as the share of fossil fuels in electricity generation declines. Implementation of CCS/CCUS technology still needs to be resolved, among other issues related to public acceptance of the technology. However, CCS deployment is still necessary, especially in energy-intensive sectors and for the emission-free production of hydrogen from natural gas. CCS will also be required to capture and store CO_2 emitted by power plants and industrial plants using biomass to achieve negative emissions.

The Paris Agreement (2015) and EU national regulations (2019/2020, 2024) set targets for reducing emissions from fossil fuel combustion, including negative emissions or CCS technology. A new financial system, a taxonomy-based crediting system after 2021, and ETS II encourage countries to use CCS technologies through subsidies for demonstration projects or compensation. However, until the ETS price/CCS cost tipping point do not overreach €150–200/tonne, "the industry will have too limited an incentive to invest in the capture facilities required for CCS, or to sign long-term CO_2 transport and storage contracts" (Jones, 2024).

CCUS technologies, i.e., technologies for the capture, transport utilization and storage of carbon dioxide, are technologies used to reduce CO_2 emissions by removing it from the gases that enter the environment from energy and industrial processes and storing it sustainably and safely. CCS technologies are also part of the European taxonomy, confirming their essential role in the European economy. The long-term goal of achieving climate neutrality is one of the key elements of global and European climate policy. CCS technologies are one of the available solutions in this regard and are, therefore, a valuable complement to other low-carbon solutions, making it possible to reconcile the transitional use of fossil fuels and the occurrence of unavoidable process emissions in an industry with the construction of a climate-neutral economy.

CCS technologies aim to reduce the amount of CO_2 emitted into the atmosphere. These technologies are used in industrial sectors, such as cement works or steel mills and also in the energy industry, where part of the carbon dioxide emissions occurs due to the process itself and not the fuels used. The process chain – the so-called CCUS chain – has three or four main components: CO_2 capture, where carbon dioxide is separated from other gases emitted by various sources and appropriately prepared for further stages; CO_2 transport, where the captured carbon dioxide is transported – usually by pipelines – to the storage site; and CO_2 processing/recycling (optional) and geological storage, where the carbon dioxide is finally sequestered by geochemical processes deep underground in tight and roomy geological formations. CO_2 can be removed from energy processes before, after or during the fuel combustion process in the boiler.

There are three main methods for capturing carbon dioxide in organic matter combustion processes:

1 The pre-combustion method (generation of decarbonized gas, e.g., after gasification of coal, biomass, waste, etc.). Removing CO_2 before combustion, or so-called pre-combustion, is a solution based on coal gasification in the presence of oxygen under high-pressure conditions. This process produces a synthesis gas containing hydrogen and carbon dioxide.
2 Post-combustion (removal of CO_2 from post-combustion gases). Oxy-combustion can also remove carbon dioxide, i.e., fuel combustion in a mixture of oxygen and carbon dioxide. CO_2 from the post-combustion flue gas is returned to the process to regulate the combustion temperature in the boiler. The resulting flue gas is a mixture of mainly carbon dioxide and water vapor.
3 Oxyfuel combustion (use of oxygen for combustion and removal of CO_2 from flue gases containing mainly CO_2 and H_2O).

Removal of carbon dioxide after combustion, or post-combustion, is the most common method for classic coal-fired power plants. The technological system is reduced to adding equipment to the power unit, which acts as a CO_2 absorber and disrober.

CO_2 capture technologies in the cement and lime industry

More than 30 active projects worldwide involve CO_2 capture in the cement and lime industry. In Poland, too, there is interest in CCS technology in the cement sector, and by 2030 it will likely be the sector that will drive the development of this technology in Poland (especially given the withdrawal of free emission allowances in this industry). The sector is responsible for approximately 12 million tonnes of CO_2 emissions per year by 2020. In parallel to the implementation of CCS technology in the cement sector, considering, among other location issues, the lime and gypsum sector will also be subject to emission reduction processes using CCS technology – approximately 2.5 million tonnes of CO_2 per year. The amine method (MEA) and the chemical calcium loop (CaL) method are adopted as the primary methods for capturing CO_2 in implementation projects worldwide. The MEA method has been known for many years, is very effective and has become the only technology widely used in various industries. The second method is very efficient – requiring a much more significant investment and additional energy in the decarbonization process.

Geological storage of CO_2

The captured carbon dioxide must be purified, then compressed and transported to a storage site in suitable geological formations, where it is injected to effectively and permanently isolate it from the atmosphere. It is sometimes possible to store CO_2 directly near a carbon dioxide emitter. If this is impossible, the CO_2 must be transported via pipelines. In the United States alone, 50 carbon dioxide pipeline networks of more than 8,000 kilometers remain in operation. In CO_2 transport, solutions based on pipeline transport in the liquid form dominate. CO_2 can also be compressed into liquid form and transported by sea.

As far as carbon storage is concerned, according to data cited by the GC-CSI, the most significant number of storage sites (including demonstration and commercial sites) are in North America, where EOR CO_2 storage projects dominate for projects in operation. However, there are also many projects

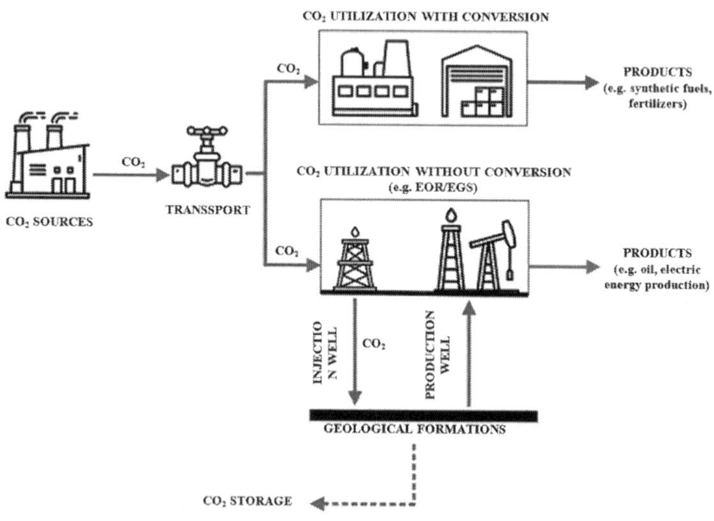

Figure 2.1 Main elements of CCS/CCUS technologies [based on: Strojny et al. 2024]

dedicated to deep aquifers. In Europe, projects related to deep aquifers predominate, and there are also projects planned to use depleted natural gas and oil fields as CO_2 storage. Whether technologies are ready for industrial-scale deployment is determined by the so-called Technology Readiness Level (TRL). For CCS and CCU technologies in general, it is rated at six on a nine-point scale by, among others, the UN Economic Commission for Europe (UNECE, 2022). However, commercial solutions (TRL9) exist and are available in each of the components of the CCS technology chain – for capture, amine-based absorption methods; for transport, pipeline transport; and carbon storage, EOR or deep aquifers. From the point of view of the Commercial Readiness Level (CRL) and the Social Readiness Level (SRL), specific CCS solutions are rated at the upper end of the scale, indicating an increasing openness of financial institutions to finance CCS projects as well as to implement adequate social and legal solutions.

In 2022, there were globally 30 operational facilities with a capture capacity of 42.5 Mtpa (million tons per annum), which is one thousandth of the emissions (GCCSI, 2022). "The total CO_2 capture capacity of CCS projects in development, construction and operation, was 361 Mtpa" (GCCSI, 2023, p. 6). While this represents an increase of almost 50% compared to 2022 (GCCSI, 2023, p. 6), this number needs to increase to 1.2 Gtpa in 2030 and 6.2 Gtpa in 2050 if the Net Zero Emissions scenario is to be achieved on time (IEA, 2022). There are currently 26 facilities under construction. Of these, 20 are projects with dedicated geological storage, and one is yet to be decided. Most of these projects in advanced development are planned to be operational by the mid-2020s and before 2030. In addition, 205 projects in the early development stages are also planned to be operational along a slightly longer

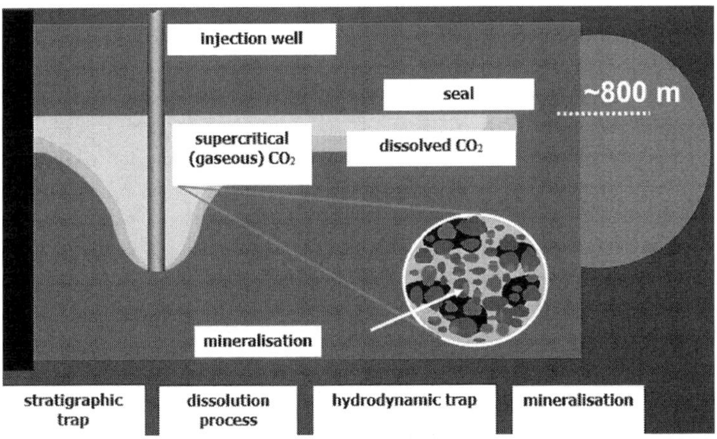

Figure 2.2 The processes involved in CO_2 storage [based on: Celia et al. 2002]

timeline (GCCSI, 2023). Although this represents a substantial increase in planned facilities and capture capacity, the IPCC (2022, 32) concludes that the current global rates of CCS deployment "are far below those in modelled pathways limiting global warming to 1.5°C or 2°C" (Langhelle, Sareen and Silvester, 2024).

Therefore, in the need for cost-effective "no-regrets" CCS value chains to be in place by that time (Langhelle, Sareen and Silvester, 2024), the EU should and does support research and activities to increase the cost competitiveness of CCS technology. CCS technology is also essential in the International Energy Agency's assessment (Technology Roadmap). CCS, in the long term, could even lead to reductions in CO_2 emissions produced by burning coal/biomass/gas. It is believed that the future of this technology depends primarily on CO_2 transport and storage infrastructure with shared access and control of CO2 transmission tariffs. "The basis for all this is to build a European CO_2 transport infrastructure that connects emitters across Europe to storage facilities," – reads the position of the Zero Emissions Platform.

The European Union is ready in principle to provide subsidies for an industrial CO_2 storage project in Antwerp, Ghent, Zeeland and Rotterdam under the North Sea. The European Parliament approved it on 12 February 2020, and declared it a "project of common interest" (PCI). In Norway, the Northern Lights project in Øygarden is completed (Phase I) and ready for injection process (2024).

Conclusions

CCS technology is important for reducing CO_2 emissions in the coming years. Among other things, CCS/CCUS technology enables the reduction of CO_2 emissions from fossil fuels and supports the fulfilment of legal and international obligations of European Union member states.

These ecological and economic considerations were the basis for a research-pilot project for CO_2 injection into the Jurassic/aquifer (AGaStor project) in northwestern Poland. This project is being prepared in the full technological chain – capture, compression and transport + storage + monitoring. An element of the project is also to determine public acceptance of CCS technology and the innovative UGS solutions with CO_2 studied in the project. The research presented in the book is the result of a team working on advancing the political and public awareness area as part of the AGaStor project.

The application of CCS technologies in Poland will make it possible to reduce carbon dioxide emissions into the atmosphere in real terms, in line with global targets to reduce the impact on climate change. CCS technologies are also part of the European taxonomy, which confirms their essential role in the European economy. The deployment of CO_2 capture and storage technologies will also be a viable tool for selected economic sectors such as commercial power generation, the cement, metallurgy, refining and chemical industries, which, on the one hand, will allow the sustainable use of hydrocarbon fuels while protecting the climate, or will be the only available alternative with carbon dioxide emissions independent of the fuels used. Thus, it will also impact the country's energy and material security, ensuring economic viability on national, European and global markets. The CCS sector will also constitute a new branch of industry in Poland, which will not only enable other sectors to function and prevent energy-intensive industries from fleeing to other countries but will also provide a base for new jobs and further innovation. It will also allow the economic sector of CO_2 utilization for several industrial processes to develop, including, inter alia, the production of synthetic fuels (so-called CCU, or carbon capture and utilization) or so-called negative emissions technologies (so-called NETs).

Implementation of CO_2 capture and storage technologies will also be a viable tool for selected sectors of the economy, such as commercial power generation, the cement, metallurgy, refining and chemical industries, which, on the one hand, will allow the sustainable use of hydrocarbon fuels while protecting the climate, or will be the only available alternative for carbon dioxide emissions that are independent of the fuels used.

Thus, CCUS will impact the country's energy and material security, ensuring economic viability on national, European and global markets, making CCUS a possible new branch of industry in Poland.

3 Energy-safety Policy and the Civil Society in the Baltic Sea Region

What is energy security, and how is it structured?

Understanding energy security is only seemingly simple because we usually focus on the issue of energy supply and satisfying the needs of various entities in this respect, from individual households to extensive industrial facilities. The balance between demand and supply is the essential challenge faced by those who produce energy, those who deal with distribution and end users who consume energy for various purposes. From the perspective of three main actors of energy production and consumption: producers, distributors and recipients, we are dealing with complex relationships between them and other participants in production and regulatory processes, which means that energy security focuses on several fundamental economic, social and political processes.

The complexity of the processes constituting the system ensuring energy security causes various actors to be involved in them:

- Economic entities – corporations and private enterprises dealing with the production of electricity and heat, operating networks supplying electrical and heat to the final recipients;
- National public authorities – central governments regulating the energy market and operating the vital elements of critical infrastructure;
- Regional and local public authorities – operating in the field of local electricity and heat production, regulating investment processes in the field of renewable energy sources, providing support for electricity and heat recipients;
- NGOs and local communities – dealing with actions for sustainable investments in energy production and programs for changes in energy and heat consumption formulas.

How relationships are arranged between these entities and their roles in the whole system determines the final results of the system's functioning, ensuring stable energy security or the occurrence of uncertainty in securing the

DOI: 10.4324/9781003607151-4

energy needs of the economy and society, or even permanent energy deficit. There are four groups of factors that affect the dynamics of energy security:

- Physical factors include possession or easy access to fossil fuel resources and geographical location, ensuring favorable conditions for renewable energy sources (solar, wind, water);
- Costs – high or low energy production prices depending on access to energy resources; the level of profitability of energy prices determined by the level of energy demand, the conditions of the economy's competitiveness and the financial capacity of households;
- Technology – using traditional technologies vs innovative technologies that reduce the high risk of posing threats to the environment; the possibility of exploiting new energy sources and making them more economically viable and productive;
- Political factors – the state of international cooperation in facing energy issues. The threat of international conflicts disrupting energy supply and destroying energy sources.

The interaction of these factors determines the final shape of a given country's or subregions' energy supply system. Even with a favorable system of physical or cost factors, the stability of the energy system can be undermined due to the negative impact of the political situation (e.g., the war in the region or countries supplying fuel). There are also situations when using outdated technologies causes such significant environmental damage that the production of energy based on it loses its sense and economic justification. In such cases, we are dealing with energy insecurity, which means either an interruption to energy availability or very high energy prices that make it unaffordable. Energy insecurity usually seriously affects people's quality of life and enterprises' economic performance. Physical energy insecurity can cause severe risks to health when, for example, a lack of or incorrect heating effect may cause long-lasting thermal discomfort or even trigger some dangerous diseases. Power failures and an unstable energy supply may cause tremendous losses in the production of goods and the collapse of services. Economic energy insecurity means high energy costs that are a heavy financial burden on the whole economy and society. It is especially harmful to medium and small enterprises and low-income households. Energy insecurity also reveals the importance of social behavior, human habits, and common ways of dealing with difficulties in accessing the appropriate level of energy supply. People's attitudes may deepen problems related to energy uncertainty, increasing the risk of adverse health and environmental effects or threats to the physical safety of people and their habitats. Wrong strategies used to deal with energy insecurity, adopted by public authorities, regional and local communities and individual households, may lead to an increasing risk of deepening crisis and disintegration of the energy system.

Finding a proper balance between the physical availability of energy resources, affordable prices for their delivery to consumers and pro-saving consumption patterns is a crucial challenge for every economy and society's political system.

To better understand energy security, the most critical dimensions of this system should be highlighted. These dimensions – described below – determine the dynamics of this system and the importance of processes occurring in it for all stakeholders. The four dimensions of energy security are as follows (Guevara, 2024):

- Availability. Energy security requires relatively easy and permanent access to energy raw materials or renewable energy sources, ensuring adequate energy generation. This dimension of energy security is historically and geographically varied. It was based on coal processing, then oil and natural gas for decades. Nuclear energy plays a significant role in ensuring energy security in many countries. The importance of renewable energy sources is increasing. The key challenge now is diversifying energy sources and breaking fossil fuel dependence.
- Accessibility. Ensuring stable and adequate energy supplies for all those who need them requires access to uninterrupted fossil fuel supplies and energy production capacities adequate to demand (proper technologies, appropriately sized production installations and a fittingly sized pool of qualified workers). In the contemporary world, the key barriers to energy supply are set by growing problems with access to fossil fuel due to political turmoil (including armed conflicts and civil wars) in the regions where the richest resources are located. Another problem is related to the high costs of developing alternative technologies for energy production.
- Acceptability. Energy production has many negative environmental effects, including air and water pollution and soil degradation. Adapting energy production to environmental regulations and norms becomes a challenge. The limits of the admissibility of the negative impact of energy production on the environment are determined by society's attitudes toward this problem and the resulting political consensus reflected in the law.
- Affordability. Energy prices are exposed to a high degree of volatility due to their dependence on oil and natural gas prices. These latter show great instability. Geopolitical crises caused by political and armed conflicts in regions such as the Middle East or Eastern Europe usually result in solid soaring in fossil fuel prices, which in turn cause energy prices to achieve extremely high values. Turbulence in the energy market undermines social cohesion because it may extend energy exclusion, an existential threat for low-income households. Therefore, guaranteeing energy affordability for a maximum number of recipients is a critical priority for all public authorities.

These four dimensions create a space for analyzing the issue of energy security in terms of mutual connections between the system's key components and assessing the level and durability of the energy security achieved (Cherp and Jewell, 2014). Considering the experience of recent decades, the most significant challenges related to maintaining stable energy security are generated by restrictions resulting from environmental protection requirements and by increasing energy prices powered by the rising costs of obtaining fossil fuels. Both factors – constituting the main components of acceptability and affordability – cause strong economic and public policy tensions and stimulate numerous social conflicts.

These long-term and recurrent problems require strategic actions to deal with challenges when the threats are intensified. Ensured energy security is undoubtedly one of the most important public issues today. Its skillful solution is a factor that determines the efficient operation of the economy, the satisfactory state of citizens' well-being and the successful prospects of socio-economic development of entire countries, individual regions and local communities. The answers to these problems are, on the one hand, regulating the energy market and, on the other hand, research and development aimed at creating new, innovative technologies enabling more effective use of traditional energy sources or the discovery of new sources.

In this context, it is necessary to point to the significant differentiation of the roles of the actors participating in the energy market. While economic entities – corporations and various sizes of enterprises – play a fundamental role in ensuring the availability of energy sources, in terms of acceptability of its production and distribution conditions, as well as affordability, the state is of crucial importance, the state, its regulatory activities, as well as support for various activities of local communities and households aimed at ensuring equal access to energy for all people, regardless of their socio-economic status. The state plays a crucial role in funding research to develop new technologies for energy generation and find better solutions for energy savings and more efficient ways of using it. As for availability, the state does not usually engage directly with the mining of fossil fuels or transmission to processing places. However, it regulates this activity via licensing, monitoring and supervising to ensure standards are met, guarantee safety and protect the environment. Thus, the state plays a fundamental role in ensuring energy security, ensuring appropriate conditions for conducting business in the field of energy production and its transmission, regulating the energy market, and supporting vulnerable customers for whom the purchase of energy can be beyond their reach, pushing them into the sphere of energy exclusion. Regardless of the importance of other entities for various aspects of energy security, the state (more precisely, public institutions dealing with the energy market) is the most important player. It gathers key competencies in shaping the energy market, shaping economic relationships between energy sellers and buyers, and reducing the negative effects of fuel exploitation on the environment. The state is responsible for creating and maintaining the energy security system.

The question arises of what role society plays in ensuring energy security. In this case, it is not easy to find an unambiguous answer. Because the state is a crucial actor in the energy security system, it is clear that society influences its shape and functioning through elections that determine who exercises political power and obtains a ticket to manage public institutions, including those functioning in the sphere of energy. This indirect impact is rarely associated with the choice between alternative versions of energy policy. Nevertheless, nowadays, it is more and more often that energy policy questions become central issues of political competition, determining the choices made by citizens. The most spectacular example of this situation was the issue of the termination of nuclear power plants in Germany and Sweden, which defined the political agenda of the most important political parties.

However, contemporary societies can also have a more direct impact on energy security and its structural ingredients. This happens thanks to the activism of social movements that seek maximum environmental protection and create good, pro-healthy living conditions for the broadest possible social circles. Numerous NGOs press the state and directly on energy producers to reduce or completely liquidate the use of harmful technologies. This pressure in many countries, mainly in Europe, seriously changes the so-called energy mix to increase the share of pro-ecological technologies in energy production systematically. Nevertheless, the NGO's activity in shaping the awareness of energy consumers is important (Sadik-Zada and Gatto, 2023). Its effect is a change in the patterns of using electrical or heating devices, leading to significant savings and limiting energy demand. Generally speaking, common social attitudes regarding how to ensure energy security are an important context for determining priorities in this sphere and the selection of methods of their implementation. An important aspect of this process, still underestimated, is shaping social attitudes toward new energy technologies, particularly the use of renewable energy sources and unconventional energy raw materials. In this regard, we deal with many prejudices, myths and stereotypes that make it difficult and sometimes even block innovation. Breaking barriers and resistance to innovative technologies in the sphere of energy requires extensive educational activities and patiently building a sense of trust in solutions developed based on scientific achievements and modern engineering. Although the state plays a dominant role in ensuring energy security, the importance of society's activity in determining important aspects of this process is growing.

Civil society's involvement in the energy transition: different trajectories of change in the Baltic Sea region

The activity of organizations and civic and local governments in the sphere of ecological policy, climate policy and decarbonization of the economy is essential for shaping the agenda of these public policies and for ways of implementing them. Numerous studies show that where the involvement of citizens in pro-ecological organizations is high, the advancement of energy transition (measured by the share of renewable energy sources in the entire energy mix) is at a high level. Undoubtedly, bottom-up grassroots movements can be conducive to more efficient transitions toward low-carbon energy systems and establishing environmentally friendly energy generation and transmission solutions. However, we cannot ignore the other side of the coin, which manifests itself in the resistance of local communities against investments in wind farms and the construction of hydroelectric power plants or hydrogen installations. NIMBY attitudes may be widespread and motivate people to counteract innovation primarily when it comes to paying high costs of investments or bearing additional burdens either from the local public budget or from individual budgets of households, or both. Grassroots movements may have two faces: constructive for change and adverse to efforts that need to be invested in change (Karimi and Rodi, 2022).

The Baltic Sea region is significantly differentiated regarding the presence of bottom-up initiatives in energy transformation. Due to historical reasons, various forms of collective activities (first of all, cooperatives) are very popular in Denmark and Germany. Energy communities play a remarkable role in the development of renewable energy sources in these countries. In Finland, the other form of community is prominent: housing cooperatives play a significant role in promoting and implementing measures for efficient energy consumption and its rational distribution. Poland is a specific case because, on the one hand, there are a lot of grassroots initiatives regarding renewable energy and managing energy efficiency. On the other hand, central government policy still depends on investing in coal energy and other fossil fuels. The obduracy of the old approach led to the inhibition of prosumers and solar PV development in recent years. The tension between these two approaches defines Poland's energy transition dynamic. In Sweden and Norway, the energy communities do not play a significant role in energy transition. This is usually explained by the industrial history of these countries, which shaped centralized energy production and distribution systems that resulted in a high level of decarbonization of energy systems. Many analysts point out cultural factors that intervene in this process, i.e., individualism, which is deeply embedded in the value systems of Swedes and Norwegians. In both cases, this cultural feature makes people more responsible for measures addressing persons as individual consumers (Karimi and Rodi, 2022). This means meeting ambitious energy transition goals is possible using only financial tools and central management instruments. As for the Baltic States – Estonia, Latvia and Lithuania – there are inherited central government energy management systems. These countries are at the very early stage of developing grassroots initiatives related to energy generation and promoting energy efficiency schemes. Nevertheless, all these three countries adopted national strategies regarding developing renewable energy and are progressing in restructuring their energy sector (Slakaityte, Surwillo and Villumsen, 2020).

One can distinguish three forms of citizens' participation in the energetic transition. First, on the individual level, it is participation in demand response by setting demand-side flexibility. The function of this participation is to balance the fluctuation of renewable energy in the energy system. Finland is a country that has widely implemented such solutions by installing smart meters in consumer facilities. Second are prosumers who are individual consumers who also produce energy (usually electricity). Prosumers have a storage capacity and can contribute to flexibility in the energy system. Third is community engagement, which occurs in the form of various initiatives. The cooperation between local governments, urban activists, local entrepreneurs and scientists in the form of energy clusters became very popular and influential in Poland. These clusters are tools to implement social innovations, allowing people, small- and medium-sized enterprises and local public institutions to adapt new technologies to their needs. They are crucial to the transition of the energy system because its pace has recently slowed down due to central government policy. Various legal and financial instruments should support and stimulate all forms of citizens' participation and community engagement because, in several countries (Sweden, Norway, Poland), top-down policies and strategies are still dominating, resulting in the exposure of energy transition to the volatile political situation at the national level (Karimi and Rodi, 2022).

A political shift on the national level may turn the path of energy transition in the opposite direction; therefore, it is critical to have stable and continuous support at the local and regional levels. There are examples of backward actions that either slow the transition or even reverse it. In 2015–2023, the Law and Justice party government practically blocked the development of solar installation and construction of wind farms in Poland. This government has been reluctant to decarbonize the energy industry because coal mining has been a vital branch of the national economy for many decades. In the opinion of right-wing political leaders, it is still and will be a foundation of national security. Polish experience shows that such a top-down policy has limitations that can restrain its influence. These limitations are embedded in a profoundly decentralized public administration system. Consequently, autonomous regional and local authorities generate many initiatives to increase the share of renewable energy sources in the energy mix and advance energy efficiency. While national energy policy aims to conserve the existing energy mix based on coal, the transition process has been supported at the subnational level.

The importance of regional and local governments for the effective energy transition is emphasized in the strategic EU documents, including The European Green Deal and Regulation on the Governance of the Energy Union and Climate Action (EU 2018/1999 of 11 December 2018). During the 136th plenary session of The European Committee of the Regions (7–9 October

2019), it was stated that local and regional authorities are the best platform to engage their communities by drawing private investors and undertaking well-prepared activities to provide sustainable energy transformation at local and regional levels. Local government institutions have a chance to play an essential role in this process as administrators, suppliers and buyers of energy services (Dziekoński, 2020).

For example, in Poland, the regional governments (Provinces) have legal competencies and financial resources (mainly thanks to the EU funds) to develop the regional energetic policy and implement several measures setting out energy supply mechanisms and efficiency. Province's authorities are entitled to adopt local laws that forbid using energy materials for heating houses and public buildings (the ban usually refers to low-quality coal, wood and some fixed fuels derived from coal). Many of them adopted relevant regulations recently. These types of regulations significantly impact the reduction of CO_2 emissions and the elimination of carcinogenic dust from the air. Regional and local authorities can play a significant role in rationalizing and reducing energy consumption in municipalities; activating local activities to produce energy from local resources; facilitating the development of prosumers, energy clusters and energy cooperatives; assessing cogeneration potential and extending its usage. Regional and local budgets can participate in the co-financing energy transformation, for instance, by sponsoring building insulation or replacing heating devices. Some local governments in Poland are involved in investments in geothermal facilities and effectively use them to provide communities with heat and warm water.

The other type of innovative action in Poland is setting up an energy cluster, i.e., an agreement between various local entities regarding the production and balancing of demand, distribution or trade in energy. A growing number of clusters may become significant players in the energy market in a significant part of the country's territory, establishing diffused energy networks. It is necessary to eradicate legislative barriers and the unclear, volatile attitude of the central government to intensify the development of energy clusters and other forms of diffused energy networks. They already have a noticeable influence on the decarbonization of local economies, but their potential impact might be many times bigger. Clusters may also strengthen equity within the energy sector because they could become alternative delivery services to big energy businesses.

Energy cooperatives are the other form of enabling citizens to take ownership of the energy transition, manage it and push toward favorable cooperation in production (mainly by developing renewable energy generation) and sustainable distribution among stakeholders by spreading efficient ways of energy consumption. Cooperatives as a form of energy community are most popular in Denmark and Germany, countries with long-lasting traditions of organizing cooperation between producers and consumers in such a framework. The specific form of energy community is a housing cooperative in Finland, dealing with energy distribution and ensuring the most efficient usage.

Civic involvement is an inevitable element of the successful implementation of energy transition policies because it can lead to obtaining social acceptance for implemented measures. Still, it can trigger disapproval and active resistance. The latter may happen when people are concerned with side effects of renewable energy installations like noise and sight disturbance, nature degradation, adverse impacts on wildlife and losses in property value. It is rational to offer some incentives for communities and intensively spread reliable information on new technologies to ensure that social barriers hampering changes will be overcome. The perception of changes is an influential factor that determines their possible success or failure. It means that preoccupations for successful social innovation are applied to meticulous information strategies, providing communities and individuals with knowledge of the meaning, range and consequences of undertaken measures. The essence of the challenge is well summarized in this remark made by Michael Mehling, Deputy Director, MIT Centre for Energy and Environmental Policy Research, Cambridge, Massachusetts:

> We already know that decarbonising our economies at the pace and scale called for by climate science will face unprecedented obstacles. Increasingly, however, we are discovering that the greatest obstacle may not be a dearth of technological solutions or finance but one of public opinion, whether expressed in the form of localised resistance against renewable energy projects or rejection of national climate action by a wavering electorate, lacking social acceptance threatens to undermine progress towards the necessary energy transition.
>
> (Karimi and Rodi, 2022, i)

As far as gaining people's engagement in the energy transformation, the critical requirement to make it happen is to reach out to all potential stakeholders (prosumers, communities, municipal authorities, local entrepreneurs, local environmental activists and local civil society organizations) with exhaustive and reliable information that is not only embedded in current scientific knowledge but also addresses convincingly the widespread concerns related to possible adverse effects of changes. One of the most important contemporary questions connected to climate change mitigation and energy transition is translating the scientific and political consensus about the need to limit the human impact on our environment into societal consensus and support.

Conclusion: between geopolitical pressure and social innovations

The Baltic Sea region is a very instructive laboratory for energy transformation. It comprises countries with diverse political and economic histories, different socio-political and cultural profiles, various legal systems and energy profiles. However, these countries must respond to the same global climate change challenges and maintain their energy security. Observing how individual countries in the region respond to the challenges of energy transformation allows for better prediction of the future of the energy sector in the region and determining the chances of achieving sustainable development goals.

All countries are confronted with the geopolitical pressure implied by Russia's aggression on Ukraine. The EU's response to the war was to cut off contracts with Russia's companies for gas and oil delivery. Breaking off dependency on Russian fossil fuels pushed all countries (except Norway) to searching alternative sources of supply. This situation forced the region's countries to accelerate green and nuclear energy development activities. The paths of energy transition chosen by individual countries differ depending on their industrial history, traditions regarding public management (centralized vs decentralized energy management systems) and physical environmental conditions (possibility of building hydroelectric power plants, wind farms, offshore wind farms).

The response to the energy crisis consists of three components:

- Developing nuclear energy;
- Developing various renewable energy sources;
- Developing interconnectedness of infrastructure and synchronizing countries' policies

The general feature of all activities related to restructuring the energy sector in the region is their pro-ecological character. They facilitate installing low-carbon energy solutions and will speed up the energy transition by replacing fossil fuel generation. Geopolitical pressure becomes a broker for profound change in the energy sector toward green energy. The volatile geopolitical situation facilitates the rise of public awareness regarding the necessity of restructuring the energy sector, and it may stimulate social acceptance of changes, even if they can increase the financial burden for communities and households (Slakaityte, Surwillo and Villumsen, 2022).

The other dimension of civic engagement in the energy transition is decentralizing the energy system. In some countries, like Denmark and Germany, the decentralized energy systems are part of their tradition and are well adapted to contemporary challenges. In some other countries, like Poland and Sweden, we are witnessing the local governments' growing role in promoting new energy generation and distribution solutions. Local governments are mainly involved in such forms of energy generation as wind and solar farms and in initiatives to promote energy efficiency. They are also instrumental in empowering citizens within the processes of energy transformation.

The Baltic Sea region generally has the forerunner status in an energetic transition. Civil societies are broadly involved in this process in most countries. Despite many obstacles to the smooth implementation of energy transformation, this region is moving toward green energy fast and continuously. Civil society's involvement reinforces this process significantly. Therefore, effective and comprehensive communication with various civil society stakeholders is crucial for accomplishing the goals of energy transition and ensuring energy security of the whole region.

The success of energy depends, among other factors, on active behavioral support for the transition of the energy system. Civil society organizations can become influential agents of change, especially when they set a platform for cooperation among local environmental and urban activists, representatives of local communities and scientists representing technical competencies. There are spectacular cases of such initiatives; for example, living labs in Poland are applied as a tool to revitalize less-developed, rural parts of the country (Karimi and Rodi, 2022, 155–85). There are many advantages of involving various civil society stakeholders. Still, the most important is the authentic possibility of transferring knowledge and experience to a broader social milieu and building social capital to strengthen support for social innovations. Civil society's involvement in the energy transition is inevitably connected with the empowerment of citizens. People are entitled to take part in the decision-making process. It allows them to ensure that such values as equity and equality are respected in the new construction of energy systems. The partnership with civil society representatives provides an opportunity to take control of how the new technologies are applied and how they are adapted to local needs. This will enhance the trust of those responsible for implementing new solutions and confirm the legitimacy of changes. Such mechanisms are necessary to obtain a proper momentum for change.

4 Polish Public Discourse on CCUS

Analysis of the Official Documents and Media Discourse

Public discourse – theoretical assumptions and Polish specificity

Public discourse is one of the key institutional mechanisms of social adaptation to dynamic social changes in terms of scientific progress, innovation and related political, economic and social challenges. The increasing diversification and professionalization of areas of social life mean that the full understanding of a given phenomenon, especially scientific novelty, is becoming reserved for increasingly narrow categories of specialists, while the potential significance of this phenomenon remains important for a much larger number of stakeholders. However, including a specific complex problem or challenge in the process of collective discussion does not mean using only rational and logical argumentation based on the best possible state of knowledge at a given moment.

> Even though more scientifically literate people might read relevant columns in newspapers or consult professional literature on the topic in order to attain a more accurate understanding by scientific standards, the majority of people do not possess the necessary educational resources or the time necessary for such research.
>
> (Wagner, Kronberger and Seifert 2002, 323)

In a contemporary society, every novelty surpassing everyday experiences that is the subject and basis of political and economic decision is being mediated within inter-personal and mass media communication. Such novelties include scientific and technological innovations that constitute a response to key civilizational challenges and are implemented as a result of decisions made by central authorities in specific local social contexts, often using the argument of necessity in the name of the "common good". However, "the less proximal a phenomenon is to [people's] immediate life space, the more they depend on media communication and less they can rely on personal communication. Mediated communication dominates personal communication in the case of

DOI: 10.4324/9781003607151-5

scientific, technological and economic innovations" (Wagner, Kronberger and Seifert 2002, 324). Communication is not a linear process by any means – it connects different threads and subjects by shared contents that are mediated in emotional and context-related dimensions (Vercelli et al., 2013). Constituting knowledge, social practices and power relations grounded on accepted visions of reality (i.e. constituting both conscious and unconscious social basis of attitudes and actions) is called a discourse (Foucault, 1977).

> Discourses can be seen as dynamic processes forming and changing our perception of reality, based on particular ways to speak about and understand the world. Accordingly, discourses act as differing frameworks for interpretation. Participants in a particular discourse will describe "reality" based on shared ideas and explanations. Another way to understand discourse is thus as mental and linguistic patterns organizing the world and adding significance to human experience
>
> (Røyrvik, Schei Olsen and Aasen 2012, 286).

By every means it is the process of simplifying the complex reality and attempts to relate the novelty to familiar everyday contexts – gathering as little information on an issue as necessary to make judgments and decisions is natural for human beings. That is why the constructed meaning(s) refer, by subtle visual and terminological features, to rather unconscious assumptions or schemata, shaping opinion on heuristic information (Ashworth and Quezada, 2011, 6195). The simplifications of a complex reality that make the discourse effective are "frames" – structures of beliefs, appreciations and prejudices that organize experience (Gunderson, Stuart and Petersen, 2020). "Ongoing media and personal communication places divergent interpretations in competition with each other. Some are eliminated, and others are embraced by a majority of people, and the group attains a new conventionalized interpretation. It has a structure that captures the complex of ideas in an image-like schema and weaves it into the fabric of the group's common-sense. This may take the form of metaphor, image, interrelated beliefs or of any other symbolic representation that can serve as an interpretation of the new. Which image is accepted in a group is a problem neither of 'scientific truth' nor of arbitrary choice, but rather is determined by a group's world of experience. That is, the resulting image or metaphor does not need to be 'correct' or 'accurate', but plausible" (Wagner, Kronberger and Seifert, 2002, 326). It has nothing to do with neutrality or objectivity. Framing an important issue means an arbitrary process of promoting a particular definition and casual interpretation along with resignation of other definitions and interpretations in order to make chosen aspects of reality more salient (Gunderson, Stuart and Petersen, 2020).

The above assumptions about theoretical foundations of public discourse and framing of the technological innovation lead to the question about the key participants in this process. The first associations lead to pointing primarily to specialists, experts and scientists who develop a specific solution from time to time for the needs of government institutions. Together with legal regulators and public services, they are responsible for containing and averting the potential harm posed by a technological innovation (Wagner, Kronberger and Seifert, 2002, 324). On the other side, there are lay people who are affected by these technological innovations in the sense that they significantly affect and change their daily functioning so far. Although this list of stakeholders seems to be insufficient and simplified, it sets the basic framework for public discourse. It is also worth remembering about representatives of the industry, international and national or local non-governmental organizations, but also regional and local authorities, and finally informal local associations of opponents or supporters of a given technological innovation along with local leaders.

Figure 4.1 is an attempt to graphically present stakeholders and relations between them along with the directions of communication depending on whether it is traditional mass media or social media. On the left are located stakeholders interested in technological innovation, stakeholders responsible in contemporary society for eliminating the potential negative effects of this technological innovation are in the center, and stakeholders who are affected by the consequences of implementing the technological innovation are on the bottom and right side of the graph. Some experts and scientists, as well as non-governmental organizations (both international and local) may be directly linked to the industry interested in CCUS technologies, while others – independent – may present positions both in favor of and against this technology. The role of the central authorities is to prepare solutions that implement the assumptions of both international and national policies in a safe way while

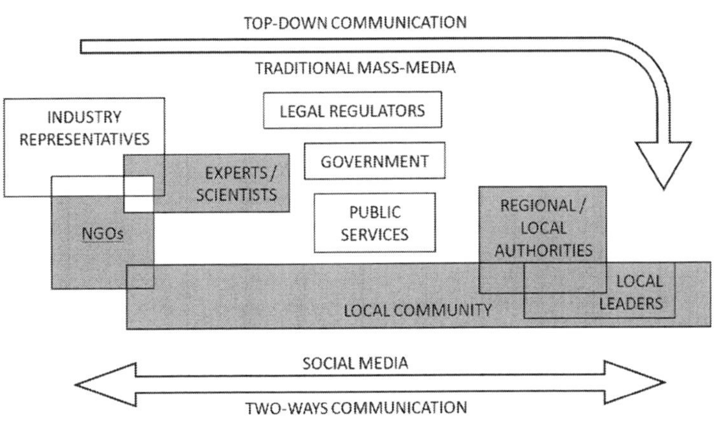

Figure 4.1 The mapping of main CCS and CCUS public discourse agents and types of communication (within the frames of central vs local contexts)

protecting the interests of the parties. The entire actual implementation takes place in specific local communities, whose formal representatives (local governments) and informal ones face the challenge of "symbolic coping" of this technology. Shaded categories of the stakeholders represent those who were the subject of research and analysis in this book.

In industry representatives' perspective, the very basis of approach toward CCUS technology is "faith in innovation" and that technology and its progress is the answer to current problems – actually, the only barrier seems to be "a matter of costs" (Gunderson, Stuart and Petersen, 2020). This approach needs instruments such as control, monetary cost, complexity reduction, effectiveness, development and efficiency (Gunderson, Stuart and Petersen, 2020). Of course, in this perspective, the lack of these instruments influences the power of innovation in a negative way. "Here, technological development and application is assumed to be inherently good as well as valuable for cost-reduction. The presumption that CCS development is good in itself can be inferred in all of the analysed documents" (Gunderson, Stuart and Petersen, 2020, 5). Both innovation and instrumental values lead to "status quo maintenance" – what in authors' opinion means a possibility to maintain current lifestyles and social structures (in consequence – decision-making process) (Gunderson, Stuart and Petersen, 2020), which directly refers to public opinion's framing of CCS. "Omissions" refer to downplaying or excluding any risks and doubts about the technology of CCS. What is characteristic for this category of stakeholders is the fact of their consistent internal compliance with the presented position within identified framings.

This cannot be stated about another stakeholders' category – experts and scientists. And it has its consequences in a more polyphonic state of public discourse.

> The fact of heterogeneous assessments and partly dissent on the CCS technology among experts strongly influence the perception and acceptance of the local population. Concerned residents find it very difficult to receive objective information and to cope with contradictory statements.
>
> (Wassermann, Schulz and Scheer, 2011, 6354)

In fact, experts and scientists actually agree only on the importance of communicating the issues to the public, the role of the state government in building a sustainable law environment and the relevance of democratic participation of all sides and stakeholders (Wassermann, Schulz and Scheer, 2011). Within all identified framings – starting from technological challenges, through law and risks and societal relevance, there is no agreement on benefits or risks.

> CCS experts, while aware of CCS risks to local communities and ecosystems in particular, see greater (particularly global) benefits. Indeed, those

> who engage directly with the public seem to be even more convinced of the benefits of CCS than those who do not. Our analysis suggests perceived CCS risks (particularly at the local level, e.g. close to a CO_2 storage site) may not be as influential in shaping expert views as they are in influencing the public, particularly local communities.
>
> (Xenias and Whitmarsh, 2018, 108)

For declared opponents of CCUS technology, the hierarchy of available solutions to the climate problem is as follows (from the most appreciated to the most unwanted): renewable energy sources, sustainable energy, CCUS technology and conventional sources, nuclear energy. In such an approach, the CCUS technology is only a fig leaf, it is used for "greenwashing" of the main cause of the climate change – conventional sources/fossil fuels, and in the same time it consumes valuable financial and knowledge resources that could be devoted to a more intensive development of renewable energy.

Polish public discourse began to take shape after 1989, simultaneously undergoing a digital transformation in parallel with Western trends. Before 1989, the mass media had operated under the information monopoly of the government, and the language dominated by the broadcaster was ineffective, because the lack of concern for the real social response meant that propaganda texts were subordinated exclusively to themselves (Bralczyk, 2007; Podemski, 2013). The beginning of the 1990s in Poland meant a dynamic development of traditional mass media (numerous new press titles, private radio and television stations appeared), which competed with each other for the recipient, at the same time consequently placing themselves in specialized niches (specific topics, selected categories of recipients) (Filas, 2010). However, already in the first decade of the 21st century, these traditional mass media were forced to adapt to the growing popularity of the Internet, soon starting a confrontation with the new online social media (Filas, 2010). The main effect of these processes is the digitization of traditional mass media and enabling online mobile access. It is worth adding that the spread of the Internet has resulted in greater transparency in the functioning of public institutions, which now publish all kinds of documents and other components of decision-making processes online. Anyone interested has the opportunity to access the most up-to-date information in this area from the source (a specific ministry, office, institution, etc.). Traditional mass media are no longer the dominant stakeholders responsible for the transparency of political decision-making processes – anyone can signal the importance of a given issue in the public space, starting a possible process of symbolically coping with it.

According to official statistical data from the Central Statistical Office, in 2023, 92.8% of households in Poland had access to broadband Internet, and on average 85.3% of recipients aged 16–74 used this access at least once a week (from 99% among the youngest to 52% among the oldest age categories) (Statistics Poland, 2023). What is specific to Poland is that almost 91% of

Polish Internet users use the Google search engine on desktop computers and as many as 98% on mobile devices (widoczni.com.pl, 2023). The dominance of this particular search engine is so significant in Poland that neologisms such as “to google” or “googling” have practically replaced the phrases like “search online” or “to look for on the Internet”.

Apart from the specifics resulting from the history of media development and consumer preferences of the domestic market, Polish public discourse fits within the content framework of the ongoing discussion on CCUS technology in the European space. In practice, this means that at one extreme are opinions about CCUS as an excuse or an alibi for industry, defending itself against the costs of transformation toward renewable energy sources. CO_2 itself is perceived as “industrial waste”, with potentially difficult to determine consequences in the long term. On the other hand, CCUS supporters emphasize its role as a mitigation tool in the medium term, enabling a gentle transition to renewable energy sources. CO_2 is treated here even as a resource enabling reduction of operating costs or even increase of profits from transport and storage.

CCS and CCUS framings in official Polish public documents and strategies

“Strategy for Responsible Development by 2020 (with the perspective by 2030)” (SRD) plays a central role as a key document for medium- and long-term economic policy (SRD, 2017, 2) of the Polish government (accepted on 14.02.2017). All other central documents are of sectoral character (all accepted in 2019: “National Strategy of Regional Development 2030”, “State Ecological Policy 2030”, “Strategy for Sustainable Development in Transport by 2030” and “Strategy for Sustainable Development of Rural Areas, Agriculture and Fisheries until 2030”), complementary and developing directions and goals from SRD. “National Energy and Climate Plan for 2021-2030. Assumptions and goals as well as policies and activities” is the document resulting from the obligation imposed by Regulation (EU) 2018/1999 of the European Parliament and of the Council of 11 December 2018 on the Governance of the Energy Union and Climate Action (NECP, 2021, 3). Of course, from the green transition perspective and the topic of Carbon Capture and Storage, the most important and most devoted to these issues are “Poland’s Energy Plan until 2040” and “National Energy and Climate Plan for 2021–2030. Assumptions and goals as well as policies and activities”. Therefore, it is possible to treat these strategic documents as governmental perspectives that set public discourse frames on green transition in Poland, giving main argumentation lines for Polish official and applicable perspectives on energy and climate. From this point of view, it is crucial to check how (if at all) CCUS exists in this argumentation.

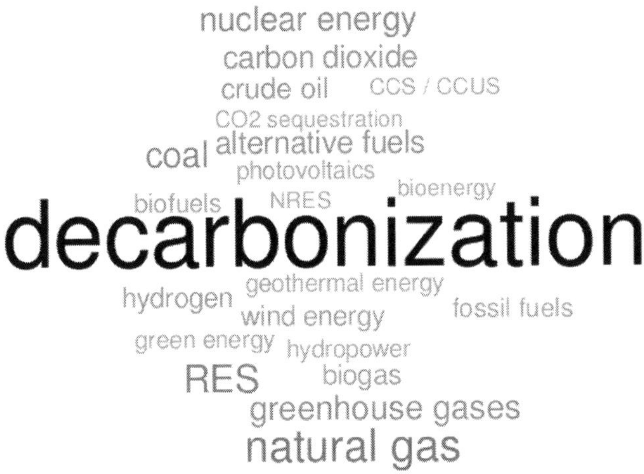

Figure 4.2 Codes' occurrences in strategic documents

If one can treat all seven state-level strategic documents as common ground for the Polish government's perspective within public discourse, then in the very center of the argumentation is, without any doubt, "decarbonization" with 36.7% of appearances in all seven texts (see Figure 4.2). Second most frequently mentioned code is "natural gas", with 11.3% of appearances. However, if all codes referring to renewable energy sources (RES) are summed up, then this issue is number two – with 18.3% of summed appearances in all analyzed government's strategies (hydrogen – 2.9%, biogas – 2.6% and wind energy – 2.4%). It must be stressed that the non-specified "Renewable Energy Sources" code occurs 286 times (8.0%) – it is significant because "nuclear energy" is mentioned 217 times (6.1%). This simple juxtaposition proves that the central goal in the governmental perspective within the discourse on green transition is "decarbonization" while the means to achieve this goal are "Renewable Energy Sources", "natural gas" and "nuclear energy". These four issues account for 72.4% of all occurrences of all 23 identified codes.

On the other hand, codes relating to sources of the problem, such as "coal" (6.26%), "crude oil" (3.7%), "fossil fuels" (1.7%) and "non-renewable energy sources" (0.4%), merely constitute together 12% of all occurrences of all 23 identified codes. And the consequences of Polish structure of energy sources (77% of the energy based on coal), such as "greenhouse gases" (5.7%) and "carbon dioxide" (nearly 4%), account for 9.7% of all appearances of 23 codes in seven documents. "CCS (and CCUS)" occurs only 10 times (of which eight times in "Poland's Energy Policy until 2040") and related code "CO_2 sequestration" appears 11 times – 8 times in "State Ecological Policy by 2030". This interesting aspect is explained in the further part of the report; however, it definitely represents governmental approach toward this issue. CCS as direct and feasible medium-term technology is only a margin of reflection on the possible ways to achieve the assumed decarbonization rates (although officially almost half of the energy in Poland will still be based on coal by 2040).

The very basis of all assumptions constructing official governmental perspective is the structure of energy production in Poland (statistical data from 2018): 77% is based on coal, 13% is from renewable energy sources, 7% is based on natural gas and 3% comes from other sources (NECP, 8). It has two major consequences – it provides an appropriate level of state energy security in connection with having its own resources (1. energy security) and mining is currently the main source of employment for around 135,000 workers (2. socio-economic importance of mine industry). These two are at the heart of the concept of "just transition".

From state energy security point of view, Polish government perceives natural gas as a "transitional fuel" (PEP, 2021, 6); however, coal will still be a significant sources of energy in next decades – according to newest "Poland's Energy Plan until 2040", less than 56% of energy will be produced from coal and at least 23% from renewable sources. The possibility of diversification of natural gas suppliers is one of the biggest advantages of this source of energy. That is why the Polish government puts that much stress on diversification, expansion of the gas network and building new gas hubs. It is possible to state that in governmental framing of green transition, "decarbonization" means reduction of energy production from coal, while in EU policy perspective this reduction of coal energy is a mean to lower greenhouse gases emissions. In other words, something perceived on an international level and within common European policy as a mean to achieve the main goal, in Polish state documents is presented as a main goal and challenge. That is why Polish strategies concentrate on social and economic costs of "decarbonization" and natural gas is presented as transitional fuel or nuclear energy is presented as an option at all.

From the political perspective – identified within the official governmental framing of the issue – the key is to achieve decarbonization indicators while taking into account the specific starting point of Poland (77% of energy production comes from coal). The change is rather "forced" by external factors – both the common EU policy and market mechanisms of increasing costs of CO_2 emissions – and the framing is concentrated on "decarbonization" itself, as the main goal. The analysis of the documents does not show that the Polish government has a more far-reaching goal of counteracting climate change and trying to adapt to new development trends in developed countries. At least such concern has not been expressed explicitly in these central level documents. What is visible, however, is the desire to adapt to the above-mentioned requirements at the lowest possible political and socio-economic costs. Therefore, among the most frequently indicated potential solutions are natural gas (as a transition fuel) and renewable energy sources with nuclear energy as target solutions.

What is ignored is the fact that, even beyond the horizon of 2040, a significant part of the energy generated in Poland will still come from coal, and this trend will most likely continue until the end of the century. Nuclear energy may provoke social protests in relation to the hierarchy of "clean" energy (not mentioning the economic costs of implementing the program on a national scale), while even doubling the share of renewable energy sources may not be simply enough to meet the growing demand for energy every year. From this perspective, CCS and CCUS technologies seem to be a proper mid-term answer to the still existing problem of CO_2 emissions in the wider context of climate change mitigation strategies. However, they are not at the center of the current government's official documents' narrative.

Polish traditional mass media and new social media CCUS framing

Between 01.01.2020 and 31.12.2023, the issue of CCS and CCUS appeared only a few times on traditional mass media in Poland – none in 2020, four times in 2021, two times in 2022 and five times in 2023.

14.05.2021 – "Gazeta Prawna" Newspaper: "Pumping CO_2 under the sea: how South Korea wants to achieve climate neutrality" – an article describing South Korean plans to reuse Donghae dried-up gas field for storing captured carbon dioxide.

19.09.2021 – "Newsweek" Weekly: "Will the great absorber save the world from climate catastrophe?" – an article on Islandic "Orca" installation for CO_2 capture directly from the atmosphere.

27.09.2021 – "Business Insider" Portal: "The world's largest CO_2 reducer is already in operation" – an article in the "News" section informing about the launch of the Icelandic CO_2 absorber from the atmosphere, called "Orca".

11.11.2021 – "Rzeczpospolita" Newspaper: "The first large factory extracting CO_2 from the air is now operating in Iceland" – an article in "Technologies/Innovations" section, presenting "Orca" installation and main technical assumptions of the operation of this installation, located near Reykjavik.

13.03.2022 – TVN Meteo TV Station: "Changing land for growing crops could significantly reduce carbon dioxide emissions" – a video material on the results of research project of international scientific team from Cambridge University (the UK), Beijing University (PRCh) and Université du

Québec (Canada) focused on potential of natural ecosystems and agricultural crops to absorb CO_2.

18.06.2022 – Polish Public Television TVP: "New CO_2 capture techniques could halt climate change" – an article in the "Science" section, describing a new type of a solvent for CO_2 capture, the use of which should be 20% cheaper than commercial technology. The technology was developed by American Pacific Northwest National Laboratory scientists.

11.05.2023 – "Rzeczpospolita" Newspaper: "Scientists have developed a new method of capturing CO_2. It is supposed to be cheap" – an article in the "Climate/Green Technologies" section, presenting the results of work of Saudi Arabia and China researchers who had managed to develop a new method of capturing and storing carbon dioxide with the use of a single-crystal caltrate salt based on guanidine sulfate. The method that should be not only cheaper than already existing but also more efficient.

15.08.2023 – "Wyborcza" Newspaper: "The United States will build two plants to capture carbon dioxide directly from the air" – reprint of an article from "New York Times", informing about the Biden administration's plans to fund two pilot projects testing CO_2 capture technology.

07.09.2023 – Polish Public Television TVP: "The factory will remove carbon dioxide from the air" – an article in the "Science" section, describing a project from Chile where an installation of capturing carbon dioxide from the atmosphere (to be processed into synthetic fuel) was to be installed in a factory.

09.10.2023 – "Kurier Szczeciński" Newspaper: "Scientists from Szczecin are taming carbon dioxide" – an article in "Economy" section, presenting the Polish-Norwegian "PhotoRed" scientific project on developing a way to process carbon dioxide into carbon monoxide, hydrogen and methane.

09.12.2023 – "Wyborcza" Newspaper: "CO_2 capture technologies have become all the rage. The question is, will they help reduce emissions?" – reprint of an article from "New York Times", presenting the discussion about further global actions on reducing the use of fossil fuels and CO_2 emissions taking place at UN COP28 in Dubai.

It should be emphasized that the topic of the Polish energy structure and heavy industry emissions is practically absent, and attention is focused on foreign, media-famous pilot projects. The formalized, top-down public discourse in the space of traditional mass media is practically silent on the subject of storing in Poland carbon dioxide captured in the country, but also abroad. Interestingly, a lot of attention was paid to Direct Air Carbon Capture technologies, which seem to be very media-friendly. At the same time, the issue of carbon capture and storage in the emission sectors of Polish industry (energy, cement production, fertilizers, etc.) did not enjoy such interest despite the growing risk of ETS cost increases and the lack of adaptation of these industries to rapid transformation. In traditional mass media, the topic of

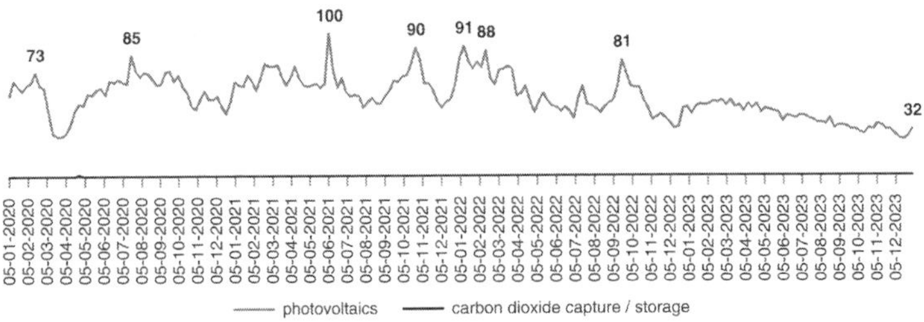

Figure 4.3 Popularity of CCS issues compared to photovoltaics in Poland in 2020–2023 according to Google search engine (Google Trends)

CO_2 capture and storage was clearly treated as a technological curiosity in the context of climate change and attempts to reduce greenhouse gas emissions.

When it comes to spontaneous social interest in the topic of carbon capture and storage, i.e., to what extent Polish residents searched for phrases related to this issue in the years 2020–2023, the results of popularity trends are shown by Google Trends (Figure 4.3). As mentioned earlier, the vast majority of Polish Internet users use the Google search engine, hence the tools provided by the operator of this search engine had been used. The search itself included two terms related to CCS – "carbon dioxide capture" and "carbon dioxide storage". The abbreviation CCS was not used because on the Polish Internet it is associated with a company providing services (the searches had been clearly geolocated to the area of operation of this company). In order to give the proper scale, both search terms were juxtaposed with the term that ignited public discourse in Poland, i.e., "photovoltaics".

The issue of "photovoltaics", which directly affected hundreds of thousands of individual households, was definitely a subject of social interest. Changes in legal regulations, restrictions introduced in government subsidy programs forced significant interest and searching for information online. The issue of reducing greenhouse gas emissions on a supra-individual scale against this background did not exist at all in Polish society. What is more, in Poland, no installation for capturing, and even less for storing, carbon dioxide has been built. No local community has had to deal with social acceptance (or not) of this solution.

The latest research on users of social media platforms in Poland (Social Media, 2023) shows that most popular social media platforms (both desktop and mobile versions) monthly average in the period from January to December of year 2022 were YouTube with 26.67 million real users per month, right before Facebook (26.37 million real users/month) and Instagram (15.88 million real users/month) (Social Media, 2023, 71). Other social media platforms, such as TikTok, Twitter, Pinterest, Snapchat, Twitch and LinkedIn, had lower reach with the age bias toward younger cohorts of users. There were no Facebook groups dedicated to CCS or CCUS in Polish language, while one could easily find hundreds of Polish Facebook groups focused on "photovoltaics" and other renewable energy sources, gathering thousands and hundreds of social media users.

Conclusions

The analysis of Polish public discourse leads to the conclusion that the issue of carbon capture and storage technology is practically absent from the public sphere. If it does appear, it is in the traditional media, where it is presented as a kind of curiosity, with an indication of foreign examples, particularly media-friendly, such as Direct Air Carbon Capture technology. In official government strategic documents, there is also a clear emphasis on the issue of "decarbonization", but as an economic and social challenge facing the Polish industry in the context of "just transition". The issue of CCS or CCUS practically does not exist in the official political perspective.

This means that in the social dimension, Poles currently do not have any "symbolic coping" strategies with technology, without which achieving the assumed climate goals in the European Union is practically impossible. Interestingly, in the case of the structure of energy sources in the Polish economy, it is CCS and CCUS technologies that may be the solution to the Polish greenhouse gas emissions issue by the end of the century.

However, it should be expected that the launch of any CCS or CCUS project will force the need to work through this issue within the public discourse, at least on a local level. However, this will mean a situation in which a given local community will only be developing specific "symbolic coping" strategies, referring to familiar, more or less fitting analogies from their own past. This is exactly what happened in the case of the described AGaStor project and the communities of Marianowo and Suliszewo in West Pomeranian Province.

5 West Pomeranian Province and Potential UGS Locations in North Poland

Introduction

The decision on localization of the research on CCUS social acceptance, as in many such cases, is primarily determined geologically. This type of research makes perfect sense in communities that may actually be adjacent to UGS in the future. The geological structure of West Pomeranian Province is one of a few that allow for CO_2 storage in aquifers. Geological research, carried out as part of the AGaStor project, indicated the Choszczno commune (Choszczno county) and the Marianowo commune (Stargard county) as the most convenient locations for UGS.

West Pomeranian Province is a part of Poland that experienced a total population exchange after World War II. This means a break in the socio-cultural continuity of the region, as well as of the localities existing in this territory today. Both in the context of historical and political conditioning of public opinion, this region, together with the Lubuskie, Dolnośląskie, Warmińsko-Mazurskie Provinces, belongs to the so-called Western and Northern Territories, which have a similar social organization both at the regional and local level (Zarycki, 2002, 72–74).

> The population formed as a result of migration, that is, mixed both genetically and culturally, is usually characterized by such features as activity, innovation, and entrepreneurship. In addition, the population of the Western Territories is much younger than the population of other large historical regions of Poland. These two factors cause the inhabitants of the Western Territories adapt faster to the new situation caused by the political change and are more susceptible to modernization processes than the citizens of other regions, especially Galicia and Congress Poland. At the same time, the inhabitants of the Western Territories demonstrate a greater capacity for self-advocacy and self-responsibility, less often – it seems – resorting to demanding attitudes.
>
> (Gorzelak and Jałowiecki, 1998, 57)

It can therefore be assumed that there is a social potential for innovation and acceptance of technologies connected with energy transition in these areas at both regional and local levels. However, as Zarycki writes, the lack of historical rootedness and "the legacy of Western and North Territories can be

DOI: 10.4324/9781003607151-6

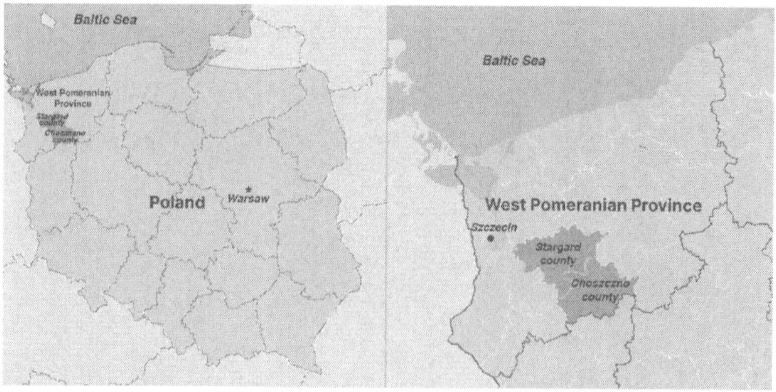

Figure 5.1 The location of the research area in Poland (survey) and in the West Pomeranian Province

seen as a context giving relatively wide room for manoeuvre to local communities" that in the existence of "the dominant role of the local factor clearly exceeding the importance of macro-regional conditions" in these regions results (Zarycki, 2002, 163).

The West Pomeranian Province is a region with a complex socio-demographic structure linked to the varying economic status of the four basic subregions. Sociological and economic criteria of division, taking into account the type of settlement (urban/rural), the type of economy (industrial/rural/tourism), and the presence of a national border, are generally close to the administrative division associated with the European reporting system (NUTS 3), in which this region is divided into four subregions. Quite accurately, these divisions overlap primarily in distinguishing a metropolitan subregion (Szczecin city) and a peripheral subregion (Szczeciniecko-pyrzycki, PL427). One planned CO_2 storage location (Choszczno) is in the last subregion. The second location (Marianowo) is just across the subregion's border, in Stargard County, administratively assigned to the Szczecin subregion. This district actually remains under the influence of the Szczecin agglomeration, with the exception, however, of the municipality of Marianowo, which has social characteristics common to the PL427 subregion. Therefore, the comparative characteristics of the districts and the municipality of Marianowo will refer to data from the province and this subregion (see Figure 5.1).

General demographic characteristics of Marianowo and Choszczno in comparison with the West Pomeranian province

The West Pomeranian Province is one with the low population density in Poland (71 persons per km^2 in 2022; Table 5.1). In the short period in which the research was carried out, a constant decrease in the number of inhabitants can

Table 5.1 Basic data about the area and population of selected localizations (counties) in 2018 and 2022 (as of 31st December)

	Area in km²	*Population total*		*per 1 km²*		*Females per 100 males*		*In non-working age per 100 persons at working age*	
		2018	*2022*	*2018*	*2022*	*2018*	*2022*	*2018*	*2022*
Province	22,897	1,701,030	1,640,622	74	71.6	105.7	106.1	64.4	71.1
urban areas	1,453	1,165,181	1,116,968	802	768.7	109.3	110	67.8	74.1
rural areas	21,444	535,849	523,654	25	24.4	98.3	98.2	57.4	64.9
PL 427	10,339	430,170	399,266	42	38.6	103.2	103.4	61.1	71.8
Marianowo	102	3,175	2,910	31	28.5	104.6	102.1	57.3	62
Choszczno county	1,328	48,602	44,820	37	33.8	100.7	101.5	60.9	71

Sources: Zachodniopomorskie... 2019; *Statistical*... 2023; Local DataBase GUS; https://bdl.stat.gov.pl/bdl/dane/podgrup/tablica

be observed in the province and researched communities. Moreover, in the rural areas of the province, the population density is more than twice lower (24 persons per km^2).

Relating these general data to local communities, Marianowo could be treated as a typical rural community. But also Choszczno county with a town capital has a density rate lower than subregion PL427. The difference between urban and rural areas in the region is also evident in the feminization rate and non-working age population rate. The feminization rate is much higher in urban areas (110) and close to 100 in the rural areas. Marianowo and Choszczno have the feminization rate slightly over 100. In the case of non-working age rate, the average for the Western Pomerania Province is 71 non-working age people per 100 working age people. The difference between urban and rural areas is very huge and took 10% during last years. Choszczno has the same rate as the average for the province, and Marianowo has the rate of non-working age people even smaller than the average for urban areas (62). During four years (2018–2022), the non-working age people rate increases in the whole province and in local communities. The provincial average increased 6.7% while in Choszczno it increased over 10% and in Marianowo only 4.7%.

Another demographic specificity of the West Pomeranian Province is one of the lowest natural increase rates in Poland which also rapidly decreased during pandemic (Table 5.2). Also the migration rate for the province is below 0. The demographic situation is little worse in the PL427 subregion but better in Stargard county (where the Marianowo community is located). Choszczno county has a huge demographic problem. On the one hand, there are the lowest rates of natural increase and migration compared with the region and sub-region. But also important is the fact that the decrease of the natural increase rate is more rapid there than in the other parts of the province between 2018 and 2022.

The age structure of the population is quite similar everywhere. The number of older people (over 65 years old) is slightly lower in two counties than

Table 5.2 Basic demographic data for Stargard and Choszczno counties in West Pomeranian in 2018 and 2022 (as of 31st December)

	Person per 1 km	*Natural increase (per 1000 persons)*		*Total migration (per 1000 persons)*		*Registered unemployment rate in %*		*Age group >65 in %*	
Year		*2018*	*2022*	*2018*	*2022*	*2018*	*2022*	*2018*	*2022*
Province	71.6	−1.9	−5.7	−0.6	−0.3	7.4	2.3	17.9	20.6
PL427	38.6	−2.4	−6.3	−3.8	−2.9	13.1	12	16.8	20.4
Stargard	78.4	−0.4	0.0	−0.4	3.63	7.9	7.1	16.7	19.6
Choszczno	33.8	−1.9	−9.5	−5.0	−3.88	16.7	15.4	16.3	20.3

Sources: *Zachodniopomorskie* ... 2019; *Statistical Yearbook of Zachodniopomorskie Voivodship*, Szczecin 2023; Local DataBase GUS https://bdl.stat.gov.pl/bdl/dane/podgrup/tablica.

the average for the region, but the most important is the rapid increase of that age category in the whole population in the last four years. It can be said that there are similar resources in local labor markets everywhere, which slowly decrease. The demographic data describe the picture of Choszczno as a community with some social and demographic problems, and Stargard as a typical local county in that part of Poland. The economic parameters strengthen that vision of analyzed counties. The unemployment rate clearly shows the difficult situation of Choszczno, where it is over six times higher than the province average and higher than in their subregion. Four years ago, the unemployment rate in Stargard only exceeded that province average. Unfortunately, since 2023, it has decreased very slowly. Hence, one would expect that both communities would need projects and investments that could counter the above developments.

Political determinants of the formation of public opinion in Western Pomerania, Choszczno and Marianowo

The history of the region translates into political practices. The political specificity of these regions, according to Zarycki, is connected with the occurrence of the left-right axis of views, specific to Central and Eastern Europe, in Poland in the average historical period. It also overlaps with other social divisions:

> Just as the left-right axis was primarily related to the differentiation of the level of religiosity and the attitude to the socio-political role of the Catholic Church, the axis of interests was obviously related to the urban-rural divide, and more specifically: to the opposition of the liberal and well-educated intelligentsia electorate to the more traditional and statist peasant electorate. Let us note that in this conflict the discursive-elite and structural dimensions were clearly divergent. Despite the very strong opposition between the "urban" and "rural" camps, practically excluding any movement of the electorate on this axis, the confrontation between these two poles never dominated the political discourse. Already in 1990 Tadeusz Mazowiecki was seen mainly as an opponent of Lech Walesa, not Roman Bartoszcze. Similarly, throughout the next decade the "urban-liberal" camp (UD- Democratic Union, KLD – Liberal-Democratic Congress, UW Union of Freedom, UPR Real-Politics Union, etc.) were primarily engaged in polemics with the anti-communist right-wing camp, despite the much greater closeness of the structural features of the two electorates than the separateness on the urban-rural line.
>
> (Zarycki, 2002, 59)

On the scale of regional differences, the Western Territories, including the West Pomeranian Province, are shifting toward the left, liberalism, modernism and urbanism as described earlier. This is shown, for example, by the results of the 2019 parliamentary elections, which deviate from the overall national results (Table 5.3).

Table 5.3 Parliament elections in 2019 – voting in regional and local communities and the political preferences in survey 2021

Party/Votes 2019	*Poland*	*West Pomeranian*	*Stargard County*	*Marianowo Commun.*	*Choszczno County*	*Choszczno Commun.*	*Sample 2021*
Civic Coalition	27.4	35.71	32.55	20.71	29.48	35.55	26.1
Law and Justice	43.59	35.11	37.84	54.73	40.94	36.57	23.5
Democratic Left	12.56	15.25	14.02	8.3	13.64	16.57	11.8
Polish People's Party	8.55	7.4	8.53	10.36	10.36	6.16	4
Freedom and Independence Confederation	6.81	.53	7.05	5.89	5.57	5.16	10.8
Poland 2050	–	–	–	–	–	–	23.9

Sources: https://sejmsenat2019.pkw.gov.pl/sejmsenat2019/pl/wyniki/sejm/okr/41; https://sejmsenat2019.pkw.gov.pl/sejmsenat2019/pl/wyniki/sejm/okr/40; AGaStor research.

In 2021, the survey sample accurately indicated the existence of a stable predominance of "left-wing" over "right-wing" views, taking into account significant shifts in political preferences during this period (the emergence of the Poland 2050 party). But in the case of researched counties, there is a clear shift toward more right-wing parties in the sense of skepticism toward the free market, orientation toward religious values, traditionalism and rurality (these are LaJ and PPP). Going down to the level of municipalities, this traditionalist direction strengthens in the case of Marianowo and is small in the case of Choszczno because the town inhabitants dominate their village surroundings.

The political orientations of the residents of the communities where the UGS project is being implemented are important for several reasons. First, they could determine the attitude toward new technologies, which can be read from the very divergent answers to questions about acceptance of technologies that cause leakage of natural substances (implicitly, storage facilities for such substances). There are clear differences in the acceptance of this situation by supporters of the most popular parties. In the case of the Civic Coalition, such technology has 46.6% of supporters, while in the case of Law and Justice, it has only 31.6%. But political orientations involve, as political sociologists have noted, more serious worldview differences that are sometimes more enduring than a preference for one party or another.

With respect to attitudes in support of new technologies, including CCS and UGS, it has been possible to establish a scale of attitudes ranging from those who reject the possibility of changing their lifestyles to preserve the environment to those who are willing to adopt a new pro-environmental lifestyle. This is based on an index created for our study of readiness for even severe lifestyle changes due to the desire to save the environment (ranging from total reluctance and conservative attitude to full acceptance). Based on 12 questions concerning care for one's environment and climate, a five-grade index of readiness to change the lifestyle was constructed: definitely averse to greener lifestyle changes; rather averse to greener lifestyle changes; ambivalent; rather eager for greener lifestyle changes; definitely eager for greener lifestyle changes.

Figure 5.2 shows that there is evident difference between four central and "left-wing" parties in comparison with more traditional ones. Voters of Civil Platform, Poland 2050 and leftist parties are eager for greener lifestyle changes (over 60% of those categories). Voters of traditional, "right-wing" parties are rather ambivalent in perception of lifestyle changes. Most of them (except L&J) have a visible group (9–12%) which is against lifestyle changes. However, it is important to remember that the above characteristics apply to the total population of the region and not take into account the perception of municipalities interests. Thus, they inform us about what the average views are in these populations.

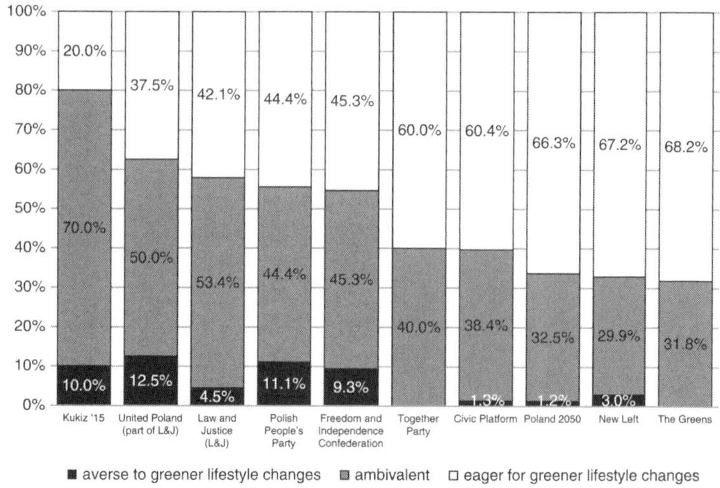

Figure 5.2 Index of readiness to change the lifestyle and political orientation in West Pomerania (2021) [AGaStor Research]

From the perspective of dissemination of knowledge about CCS and UGS technologies, it is also important to know the structure of local authorities, which should be considered independently of the political orientations described earlier. Thus, in both communes – Marianowo and Choszczno – local election committees have a very strong position, while the mayor of Marianowo (Danuta Ankutowicz, with her own representation in the commune council) and the mayor of Choszczno (Robert Adamczyk, whose committee has a majority in the commune council) are people who are not affiliated with national parties.

The issues raised in the local government campaign, which show that also the voters have a pro-investment attitude, are fundraising and cooperation with business:

> The only candidate giving Our Municipality a real chance for development. Ms. Danuta has extensive experience in the use of EU funds, management of associations and experience in working in the commune structures as well as in business.
>
> (voter opinion from Marianowo on the Internet)

In both communities, it can be seen that local election committees are partially connected with certain region-wide political preferences, but the strength of those committees is their focus on local issues.

This peculiarity of local politics was very clear already during the research. The appearance of some opposition among counselors was not at all related to worldview/political elements but to different perceptions of local issues and needs. This does not change the fact that local politicians also have preferences binding them to specific parties from the national level. In the case of two researched communities, we can recognize these orientations of politicians as related to the centrist parties of the Civic Coalition. However, this is not related to the distribution of residents' political preferences – especially in Marianowo. Even at the community level, some national parties have their own municipal representations. It is the case of Marianowo where PPP has a very strong representation. Therefore the influence of this party on the local acceptance of CCS and UGS technologies should be taken into account.

Selected economic data about West Pomeranian province

The infrastructural aspects take into account also some landscape elements. The characteristics of the landscape – and above all those elements which are socially recognized values – are important because the local communities and pro-environmental NGOs may recognize the CCS/CCUS technology as a potential threat to some of these elements. Data from the Central Statistical Office of Poland from 2018 indicate that the largest territorial areas under the legal protection are located in the Choszczno county while in locations in Stargard the size of valuable landscape territory is well below the average for the whole region (all data in Zachodniopomorskie … 2019).

However, taking into account the qualitative data describing ecological and tourist values of analyzed counties, it can be noticed that in Choszczno county most of the legally protected areas are associated with bird habitats distinguished by the project Natura 2000. The location of UGS in the sites in that county probably will not infringe the value of these areas, therefore the presence of these areas cannot be considered a significant obstacle to social acceptance for the location of the gas warehouses.

Stargard county has a small number of reserves and landscape points but a high number of valuable natural and cultural objects. Those objects are surrounded by tourist infrastructure (mainly tourist routes), which may cause the local community to recognize a conflict of interest between the activities carried out in the CCUS project and the local tourism industry. Location of the UGS facilities might be negatively perceived by local interest groups, although the trends observed in the accommodation base indicate that the tourism industry is weak in those two counties in comparison to the whole region (see Figure 5.3).

In the context of the development of an accommodation infrastructure (water supply, sewage, gas), Choszczno is characterized by an average infrastructure for the region, while Stargard has a high level of the infrastructural

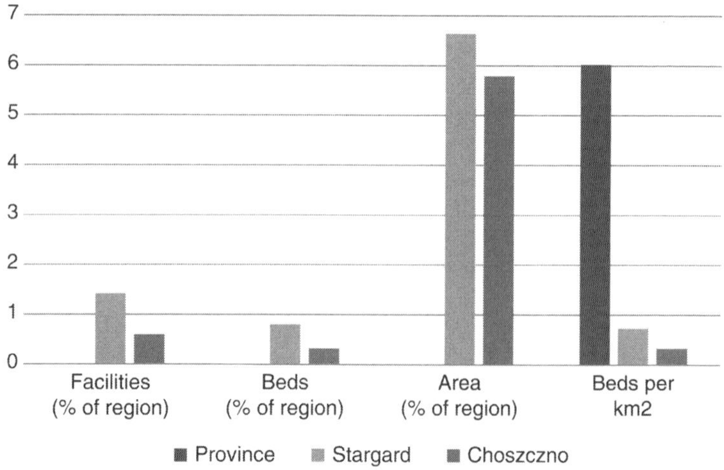

Figure 5.3 Tourist accommodation establishments in 2018 [Zachodniopomorskie … 2019]

development. But that last data is common for the relatively large town which belongs to the county and very rural surroundings with the Marianowo county, where the site for UGS could be located.

Because of high local expectations connected with the local development of the tourist market, it is necessary to focus on infrastructural potential of Marianowo and Choszczno in this branch of services. Data from 2012 on accommodation facilities in the West Pomeranian Province show that two researched counties have a very poor structure of hotel facilities (less than 1 hotel's bed per 1 km). This means that the influence of the tourist lobby is weak there, even in the whole area of Stargard county. Data from 2018 show that the development of tourism in the province is very uneven. It grew throughout the region but remained very underdeveloped in Choszczno. The number of beds in Stargard county decreased over the period of four years.

Choszczno has higher per capita income ratios than Stargard, but it also systematically gets into debt on a larger scale than Stargard. Taking into account that some local government revenues are a national government subsidy, this data can be interpreted as an indicator of the depopulation of Choszczno (because not all state subsidies are connected with the number of inhabitants of the community). Indebtedness in Choszczno does not result from the level of investments because the investments outlays in this county are more than half lower than the province average. In Stargard, the investment outlays are above regional average. It is worth noting that Choszczno was included in the Polish Academy of Science list of endangered towns in Poland in 2024 (position 81; see *Samorząd*, 2025). It was in the third risk group (losing of town's function, highly unfavorable socio-economic situation).

Data from the National Court Register show that enterprises and associations are naturally concentrated in larger centers, which is why there are much more of them in Stargard than in Choszczno (also when this number is taken into account per capita). According to data from the Central Statistical Office of Poland, the number of economic entities in those locations is lower by 25–35% in relation to the regional average (in 2018; all data from Zachodniopomorskie ... 2019). The average salary in these countries is within 81–87% of the regional average. Taking into account the higher salaries in Szczecin compared to the entire region, these figures can be considered average for the province.

Statistical characteristics present Choszczno as a demographically and economically weak area, with unused labor potential, lower social and economic activity compared to the region and to Stargard. It seems that these social features may be a factor making social acceptance of CCUS technology easier when it will be presented as pro-development investments.

Marianowo, on the other hand, was selected in 2009 as the location of the LPG warehouse (Duszczyk, 2009). Although this plan was abandoned after a few years at the design level (2012), the positive attitude of local authorities to this idea was noticeable (*Gigantyczny magazyn, 2012*). Those media information suggest the existence of a positive attitude to this type of investment in Marianowo and local self-government.

Demographic and economic determinants of attitudes toward CCUS and UGS

When analyzing the importance of social factors on individual attitudes, it is important to keep in mind both the socioeconomic vulnerability of the communities of interest and the subjective role assigned to material values in these communities. In the sample surveyed, 70% considered economic factors important to their quality of life, while 5% considered them unimportant. After declarations about health (research conducted during the COVID-19 pandemic), this was the highest rated value. Recognition of economic values significantly correlated with the index of readiness to change the lifestyle to a greener one (Pearson Chi Square .000; Likelihood Ratio .000; Linear-by-Linear Association .000; Table 5.4).

In the same way, this translates into the willingness of this significant group of people to support new technologies in general, also in situations where it involves some inconvenience, such as leakage of natural substances. However, this relationship is most apparent for those who consider the impact of material factors on their lives to be very important. They are especially likely to accept new technologies if they convey tangible benefits to

Table 5.4 Economic factors of the life quality and index of readiness to change the lifestyle

Index of readiness to change the lifestyle	*Life quality: economic factors (stable employment, savings)*					
	to a very low degree	*to a low degree*	*Moderately*	*to a high degree*	*to a very high degree*	*Total*
Definitely averse to greener lifestyle changes			0.6		0.5	0.3
Rather averse to greener lifestyle changes		3.8	1.8	3.7	3.2	3.0
Ambivalent	62.5	53.8	65.1	35.1	29.9	41.7
Rather eager for greener lifestyle changes	37.5	23.1	26.6	45.4	45.7	40.0
Definitely eager for greener lifestyle changes		19.2	5.9	15.9	20.8	15.0

Source: AGaStor research.

the county/municipality (Pearson Chi Square .000; Likelihood Ratio .000; Linear-by-Linear Association .000; Table 5.5)

Unfortunately, however, these correlations that show strong acceptance by materialistically oriented people of pro-climate sacrifices, green energy, and investment in new technologies do not translate into acceptance of CCUS and UGS technologies, mainly because these people have no knowledge, for example, that these are technologies associated with clean energy sources (47% of the strongest materialistically oriented respondents have never encountered such an opinion). Thus, it can be assumed that the factor significantly reducing the potential for acceptance of CCUS on the basis of economic benefits for local communities is the lack of association of these technologies with clean, or green, energy.

Table 5.5 Material values and acceptance for investments in new technologies with the ecological and economic benefits

Acceptance for investments in new technologies beneficial to curbing climate change if they'll provide specific economic and energy benefits to a given commune and county	*Life quality: economic factors (stable employment, savings)*					
	to a very low degree	*to a low degree*	*Moderately*	*to a high degree*	*to a very high degree*	*Total*
Definitely no			3.0	1.8	1.8	2.0
Rather no	12.5	7.7	7.1	6.6	5.0	6.3
Hard to say, I don't know	25.0	46.2	55.0	32.1	27.6	36.7
Rather yes	50.0	42.3	29.0	50.2	38.9	41.2
Definitely yes	12.5	3.8	5.9	9.2	26.7	13.8

Conclusions

The knowledge of some demographic and social characteristics of the region's population and referring to the quantitative research make it possible to notice three main factors that build the potential of social acceptance of CCS and UGS technology. These are: orientation to material values, inclination to change the lifestyle to a more ecological one and orientation to technological development.

The first element of social awareness is the most prevalent and correlates with positive attitudes toward new technologies, highly valuing the potential economic benefits associated with them.

The second factor correlates with political views but also indicates a strong innovative attitude among the region's residents. The two dominant categories on the scale of lifestyle changes are ambivalent attitude and rather accepting such a change (40% of respondents each). The attitude of definite readiness for such a change is relatively numerous (15%), while attitudes excluding such a change altogether appear in only 5%. This means that the entire population of the region – as described by researchers of worldview issues – is shifted toward acceptance of pro-environmental values.

The recognition of the development of technology as a factor determining the quality of life appears in 48% of the respondents, so it is a moderately common value. As the results of the survey will show, it is particularly important for the potential of social acceptance of CCS and UGS technology.

The analysis of political culture and demographic structure of West Pomeranian Province show that region as more "westernized" than other regions in Poland. It could be also observed in the fact that regional societies accept pro-environmental values. In the regional economy, the proximity to projects related to the energy transition, namely numerous wind farms that have sprung up over the past years (Wind Power, 2024). On the other hand, the local determinants of Choszczno and Marianowo diverge from the regional characteristic. Both levels are important for understanding the opportunities and obstacles of the realization of CCUS which are described in the next chapters of the book.

6 Uncertainty and Ignorance

The Regional Awareness of CCUS in Survey Research

Introduction

The chapter refers to the results of the quantitative research made in the West Pomeranian Province. Two potential UGS locations were recognized in geological research in that region, and the research was focused on the understanding of larger social contexts of the potential CO_2 storage location. Then the regional society which was investigated could be faced with that technology in the near future.

Regional quantitative and representative survey research identify a wide range of territorial, economic, demographic and – last but not least – worldview determinants of social awareness and social acceptance of CCUS. This allows us to explore views specific to a certain type of community, similar to where CCUS storage sites might be located, but also to see their specificity to another type of community locally adjacent to them.

CCS social acceptance in Poland – lessons from former research

Identifying the socio-cultural contexts that determine the level of social acceptance of the CCUS is possible in survey research by gathering knowledge about the demographic and social characteristics of respondents representing different views on the CCUS, and relating these views to other opinions. The first of these conditions is possible by using random sampling, as in the two studies made in Poland before AGaStor research (Brunsting et al. 2013; Kaiser et al. 2014). However, from the experience of the predecessors, it appears that a certain barrier to recognizing the influence of these factors (independent variables) is the very high percentage of people who have not encountered the term CCS, so, as Brunsting et al. (2013) recognized, they cannot be considered as bearers of specific views about this technology.

This way of thinking seems ill-suited to the realities of today's world, where many beliefs are based on highly generalized stereotypes or even

DOI: 10.4324/9781003607151-7

ignorance (Beck, 2012, 186–8). That is, ignorance itself is an important context for social responses to technology (Beck, 2012, 173–5). This is shown by the study of Kaiser et al. (2014, 272), in which people who had not heard of the technology before were also asked about CCS. Despite this, "55% of the respondents had positive expectations of what CCS might bring to the area, and only 18% thought that a CCS project would harm the region".

These researchers examined the character of that ignorance:

> A significant number of the respondent's mentioned impacts of local CCS plans that are incorrect or unrelated to CCS technology, for example that it would reduce smog (24%), toxic waste (12%), or water pollution (9.5%). These results show that even if the respondents have heard about CCS, their knowledge about this technology is very low.
>
> (Kaiser et al., 2014, 269)

In this sense, this team's research can be considered as one that produced results in Poland by identifying the socio-cultural contexts of CCS/CCUS social acceptance, which correspond with the research from other countries (Pidgeon et al., 2012). The principle of sample representativeness was preserved, while "respondents answered several questions about their region, for example, to what extent they were satisfied or dissatisfied with their area and what they considered the most important problems they faced. Further questions focused on local CCS plans. Respondents were asked how much, if anything, they knew about carbon capture and storage plans in their area before the interview" (Kaiser et al., 2014, 269).

In the case of the Rychlicki et al. (2015), sampling was a weakness. Still, it is appreciated that the survey questionnaire was extensive enough to capture the various worldview contexts of CCS social acceptance. It contained as many as 60 questions divided into three thematic groups. The first concerned general knowledge about oil production and perceptions of the oil industry. The second referred to the impact of oil exploitation on the environment and human health. However, there were also questions about the state of the environment and its place in the respondent's life and the perceived negative effects of environmental pollution (including separate air, water, soil, fauna, flora, etc.). The third group of questions already referred to the knowledge and evaluation of CCS and CO_2-EOR technologies. This group focused on respondents' opinions on the development of new oil fields, the use of modern technologies (CCS and CO_2-EOR) in investments related to the oil industry and benefits and risks associated with these technologies. Thematically, the latter group was strongly related to the CCS public acceptance research profile described in the literature (Tcvetkov et al., 2019).

The survey by Tarkowski et al. (2014) was much more modest in content, but also thematically focused on linking knowledge about CCS with opinions on worldview issues. Respondents were asked about: general views on climate change and sources of information; perception of global warming as a threat; opinions about actions; perception of causes and effects of climate change; and actions to address climate change. Only the last batch of questionnaire questions dealt with the knowledge and impact of CCS technology on climate change. As can be seen from the way the CCS/CCUS problem was described, it was already recognized in the survey in the context of climate change.

What is the methodological lesson from former research? Both the general principles of sociological research methodology (Babbie 2013) and the results of those research indicate that survey research must be conducted on representative samples. This is primarily because social awareness of CCS/CCUS technology is very limited in Poland. A research problem that has yet to appear in such analyses is therefore not only the question of what people know about this technology. The main issues are: who has not the knowledge about CCUS and how that lack of knowledge influences its social acceptance or social rejection? It is impossible to answer these questions by conducting research locally because the local population is characterized by a flattening of socio-demographic variables due to its relatively (compared to larger territorial communities) high homogeneity. This means that effective survey research preceding qualitative data collection should be conducted at least at the regional level, on a representative sample. In the AGaStor project, this type of survey was conducted in the whole West Pomerania Province, which includes the local communities (Marianowo and Suliszewo) where the project is realized.

Also in other European countries survey research recognize lack of knowledge as its output. Otto, Sprenkeling et al. (2022, 6–7) included in the questionnaire a glossary explaining the basic concepts related to CCS (so-called Informed Questionnaire). In the natural conditions – and the case of Poland is not the exception here – social actors formulate opinions and make decisions without such knowledge. Thus, the questions about CCS and UGS were geared in our survey questionnaire to recognize pre-judgments, which are essentially stereotypical, and allow us to recognize the beliefs with which common uncertainty will be filled.

The conclusion from the analysis of Polish research on CCS social acceptance is the necessity to extend the questionnaire with world-view independent variables, such as attitude to climate, state of environment, or reception of the most important social problems. Generally, it means that the object of analysis should also include the value systems applied by the respondents (Babbie, 2013; Pidgeon et al., 2012), which may influence the CCS/CCUS social acceptance as much as other conditions. Furthermore, the generalized perceptions of new technologies were also considered important.

CCUS social acceptance in Western Poland – results of the AGaStor research

Social determinants of attitudes toward CCS and UGS

When analyzing the importance of social factors on individual attitudes, it is important to keep in mind both the socioeconomic vulnerability of the communities of interest and the subjective role assigned to material values in these communities. In the sample surveyed, 70% considered economic factors important to their quality of life, while 5% considered them unimportant. After declarations about health (research conducted during the COVID-19 pandemic), this was the highest rated value. Recognition of economic values significantly correlates with support for new technologies in general, also in situations where it involves some inconvenience, such as leakage of natural substances. However, this relationship is most apparent for those who consider the impact of material factors on their lives to be very important. They are especially likely to accept new technologies if they convey tangible benefits to the county/municipality (Pearson Chi Square .000; Likelihood Ratio .000; Linear-by-Linear Association .000)

However, these correlations show strong acceptance by materialistically oriented people of investment in new technologies – in that those which are connected with green energy – do not translate into acceptance of the concrete CCUS technology. This is because these people have no knowledge, for example, that these are technology associated with clean energy sources. Forty-seven percent of the strongest materialistically oriented respondents have never encountered such an opinion. Thus, it can be assumed that the factor that significantly reduce the potential for acceptance of CCUS on the basis of economic benefits for local communities is the lack of association of these technologies with clean, or green, energy.

The most important from the perspective of the social acceptance of CCUS technology is that part of the population that declares that technological development is very important to their quality of life. Like the previous category, they are positive about new climate-supportive technologies. For the some basic statements, this looks as described in the table which shows that such technology is accepted even if it causes some local life inconveniences (Table 6.1; Pearson Chi Square .000; Likelihood Ratio .000; for 23–24 – Linear-by-Linear Association .000; for 25 – Linear-by-Linear Association .003)

The level of acceptance for these views is higher than for the previous, materialistic orientation. Most importantly, however, this also translates into statistically significant acceptance of CCS and UGS directly (Pearson Chi Square .000; Likelihood Ratio .006; Linear-by-Linear Association .000), which declares unconditional support for such investments. Such support of UGS technology increases when such an investment is tied with material

Table 6.1 Acceptance for investments in new technologies beneficial to curbing climate change even if they can cause the leaks of substances that are neutral/harmless to the environment and people (technological development as the source of life quality)

Acceptance for investments in new technologies beneficial to curbing climate change even if:		*Influence on life quality: technological development (as the source of life quality) in %*					*Total*
		to a very low degree	*to a low degree*	*Moderately*	*to a high degree*	*to a very high degree*	
They can cause the leaks of substances that are neutral/ harmless to the environment and people	Definitely no	50.0	9.3	4.8	8.2	3.0	6.5
	Rather no	25.0	18.6	17.5	13.5	13.4	15.7
	Hard to say, I don't know		37.2	41.7	35.6	28.4	37.6
	Rather yes		30.2	29.0	32.6	26.9	30.1
	Definitely yes	25.0	4.7	7.0	10.1	28.4	10.2
Total		100	100.0	100.0	100.0	100.0	100.0

Source: AGaStor research.

benefits (Table 6.2; Pearson Chi Square .000; Likelihood Ratio .000; Linear-by-Linear Association .000).

Two important phenomena for social acceptance of CCUS can be inferred from these data. First, a general trust in technological development, clocking it as an important value for one's life, results in overt social acceptance of a particular technology – in the case of the referenced studies of CCS and UGS technologies. The second important observation is that this type of attitude

Table 6.2 Agreement on underground CO_2 storage facilities: I would agree to them if they bring material benefits (technological development as the source of life quality)

Agreement on underground CO_2 storage facilities:		*Influence on life quality: technological development (as the source of life quality) in %*					*Total*
		to a very low degree	*to a low degree*	*Moderately*	*to a high degree*	*to a very high degree*	
I would agree to them if they bring material benefits	Definitely no	25.0		2.5	1.9	1.5	2.2
	Rather no		20.9	10.2	9.4	10.4	10.5
	Hard to say, I don't know	25.0	58.1	50.6	45.7	32.8	47.3
	Rather yes		16.3	31.8	38.6	32.8	33.4
	Definitely yes	50.0	4.7	4.8	4.5	22.4	6.6
Total		100	100	100	100	100	100

is not common in the surveyed population, which means that an important action for building trust in technologies such as CCUS should be building general trust in technological development. This is because it has a stronger influence on the acceptance of CCUS compared to the orientation toward material values. Compared to a willingness to change one's lifestyle to a greener one, trust in technology has a direct impact on acceptance of CCUS and not only in the context of its attractiveness for the climate.

Lack of knowledge about the CCUS technology as the main social obstacle

In the context of the analysis of the influence of worldview's orientations, the problem of ignorance as a factor inhibiting social acceptance of CCUS technology emerged. In the case of CCUS technology, those who have some association correctly associate it with capturing CO_2 produced by industry (34%) or using filters in industry (39%). Sixty percent to 65%, however, do not have this type of knowledge. What is interesting, in small communities – villages and towns – indications of correct answers are rarer than in bigger cities. On the other hand, more people in those small settlements declare they ever heard about CCS (Figure 6.1). Probably respondents in small towns rather try to hide their ignorance than have any idea about that technology.

The same is true for UGS, where correct indications are even rarer (19% and 29% correct associations). In the case of CO_2 capture, it should be noted that correct associations are the most frequent, while incorrect associations

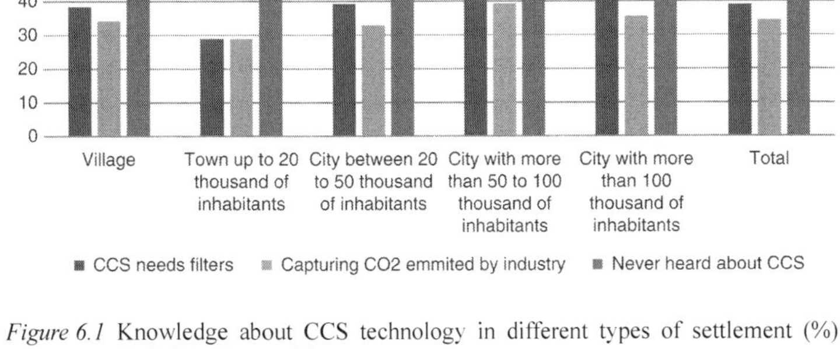

Figure 6.1 Knowledge about CCS technology in different types of settlement (%) [AGaStor Research]

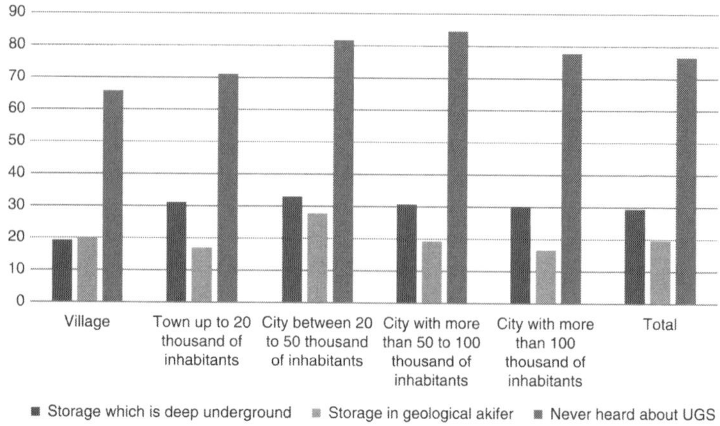

Figure 6.2 Knowledge about UGS technology in different types of settlement (%) [AGaStor Research]

(association with photosynthesis, capturing CO_2 from the air) are indicated significantly less frequently. However, in the case of underground gas storage, the scale of associations is already wider (the popular association is with natural gas deposits – 32%; Figure 6.2). This means that lack of knowledge causes a significant increase of uncertainty.

This uncertainty is revealed in the frequent "hard to say" statements. It is important to pay attention to these especially in relation to general questions about acceptance of new technologies, since previous analyses have shown that there are significant groups that are willing to accept technological developments in general. However, even with such questions there are groups that are unable to express an opinion, for example, in the question: "Under what conditions would you be willing to accept investments in new technologies that are economically beneficial to your county and municipality (new jobs, business taxes; climate taxes)". This question indicates very specific economic benefits, and when these benefits are emphasized, such a local investment receives quite a high acceptance, generally exceeding 50% of the responses (Table 6.3).

In smaller towns and cities, however, quite a large number of people (about 40%) choose the answer "hard to say". The same is true for the presentation of a relatively neutral situation, i.e., the association of the presence of such an investment with a possible leakage of neutral substances. Also in rural areas and small towns the declarations indicating uncertainty are increasing.

The next situation about which the respondents were questioned referred to the potential of a NIMBY attitude. This was the situation of separating

Table 6.3 Acceptance of local investments in a new technology and the size of the place of residence (local benefits)

Size of the place of residence:	*(If:) They will provide specific economic and energy benefits to a given commune and county (in %)*					*Total*
	Definitely yes	*Rather yes*	*Hard to say / I don't know*	*Rather no*	*Definitely no*	
Village	11.1	39.4	39.4	7.1	3.0	100
Town up to 20 thousand of inhabitants	11.1	35.6	43.0	3.7	6.7	100
City between 20 and 50 thousand of inhabitants	14.8	30.4	42.6	9.6	2.6	100
City with more than 50 to 100 thousand of inhabitants	16.2	40.0	37.7	5.4	0.8	100
City with more than 100 thousand of inhabitants	15.7	45.4	28.2	9.7	0.9	100
Total	14.1	39.1	36.8	7.3	2.6	100

a part of the commune area for an investment. Here too, the greatest uncertainty appeared in villages and small towns. This is a doubly significant phenomenon. First, it shows that regardless of the declared knowledge about CCS and UGS technologies (which varies), it is these responses that show a significant element of uncertainty and lack of opinion about new technologies in specific local investment situations. They appear in areas of the type in which UGS investment is planned in the project. Second, this uncertainty should be considered not only in the context of ignorance, but also in the context of potential resistance to investments (Table 6.4).

Table 6.4 Acceptance of local investments in a new technology and the size of the place of residence (local costs)

Size of the place of residence:	*(If:) It will involve designating a part of the commune area you live in for investment (in %)*					*Total*
	Definitely yes	*Rather yes*	*Hard to say / I don't know*	*Rather no*	*Definitely no*	
Village	5.1	27.3	52.5	7.1	8.1	100
Town up to 20 thousand of inhabitants	7.4	28.9	44.4	11.1	8.1	100
City between 20 and 50 thousand of inhabitants	12.2	24.3	40.0	15.7	7.8	100
City with more than 50 to 100 thousand of inhabitants	6.9	36.9	40.0	13.8	2.3	100
City with more than 100 thousand of inhabitants	6.5	39.8	38.0	13.0	2.8	100
Total	7.5	32.8	42.0	12.4	5.3	100

Table 6.5 Agreement on underground CO_2 storage facilities and size of the place of residence (rejection of a local investment)

Size of the place of residence:	*Under no circumstances would I want such technologies in my county, even if they are safe (in %)*					*Total*
	Definitely yes	*Rather yes*	*Hard to say / I don't know*	*Rather no*	*Definitely no*	
Village	1.0	10.1	59.6	25.3	4.0	100
Town up to 20 thousand of inhabitants	3.0	10.4	57.0	25.2	4.4	100
City between 20 and 50 thousand of inhabitants	4.3	13.0	54.8	16.5	11.3	100
City with more than 50 to 100 thousand of inhabitants	4.6	9.2	52.3	24.6	9.2	100
City with more than 100 thousand of inhabitants	3.2	9.3	54.2	24.1	9.3	100
Total	3.3	10.2	55.3	23.3	7.9	100

Analysis of the data on general acceptance of the technology shows that the problem of this type of resistance is more serious in medium-sized localities – above 20,000 and 50,000 inhabitants. These are the localities where greater knowledge about underground gas storage is declared, indicating that these communities are most susceptible to CCUS rejection. Rural and small-town communities have a lower presence of such views, which distinguishes them from somewhat larger communities. There is also a large group of people who reject NIMBY attitudes (about 30%; Table 6.5)

It seems, however, that the lack of decisiveness appearing in such answers in the range of 55–59% is a very big potential in terms of inactive resistance against local investments, which in the situation of a concrete investment proposal and the already indicated lack of knowledge may quickly become active rejection of a new technology.

Conclusions

If we compare surveys conducted in Poland a decade ago with data from AGaStor, it turns out that virtually nothing has changed in social awareness of CCS in Poland so far. Poland is that European country where the huge lack of environmental awareness is observed (Cynk, 2022). However, if we focused on the CCUS technology, it should also be kept in mind that other societies in Europe and around the world are characterized by a similar state of ignorance (cf. Ha-Duong et al., 2009; Li et al., 2014; Sitinjak et al., 2023). Thus the technological ignorance observed in the research represents the crucial recent problem of the "risk culture" and "constructed uncertainty" (Beck, 2012, 187–8).

In studies devoted to society and technology, the category of uncertainty has many meanings (Beck, 2012; Nicolas et al., 2021; Nowotny et al., 2001). However, it always functions in a categorial framework together with risk estimation, ignorance and decisions making. Many studies focus on uncertainty in scientific discourse (here the best example is climate uncertainty; Nicolas et al., 2021) or in the implementation of new technologies (Otto, Sprenkeling et al., 2022). In the context of the latter considerations, opinion polls are also being conducted to test the extent to which ignorance is socially accepted in the context of the estimated risks of some technologies (in that CCUS; Otto, Pfeiffer et al., 2022)

Referring to Beck, this type of ignorance is called reflexive. Also Otto and Gross (2021) notes that the communication of such ignorance, which heavily distorts the reception of technology, depends on communication styles. In contrast, other research shows that new social technologies are recognized as un-resolved or long-term uncertainties (so-called unknown-unknowns; Pidgeon et al., 2012, 4181). Pidgeon noted also that the opinion/decision-making process involves other competencies of the respondents involving values and trust:

> The study of public perceptions and beliefs can provide insights into value and ethical concerns not ordinarily considered by scientists and engineers, or factors such as trust in risk management, all of which are important for the governance and control of risks but which often prove more difficult to express in formal risk–benefit assessments. Perceptions also matter because *beliefs (even when they turn out to be completely wrong) can have real consequences in the world.*
>
> (Pidgeon et al., 2012, 4177)

So, the uncertainty in the context of survey research has a slightly different meaning. It is important to realize that it is not just technological or financial uncertainty (L'Orange Seigo et al., 2014) but strictly social. In quantitative research, it can only be assumed that uncertainty refers to the need to make a decision (including giving an opinion) in a situation of limited/lack of knowledge about a topic. Thus, with regard to studies that do not involve experts and are not in-depth qualitative studies, it should be assumed that they reflect all forms of ignorance: reflexive one, known-unknowns, unknown-unknowns and reluctance to obtain known (Beck, 2012).

The analysis of the opinions of respondents without knowledge of CCUS technology thus reveals how different social groups manage their ignorance and uncertainty, which is very relevant in the context of contemporary risk staging, as described by Beck (2012). This is because limited knowledge of potential risks and hazards does not obviate the need for decision-making on a broad social scale, in which we are dealing with political decisions (understood not only as decisions of politicians, but also of citizens – including both support for certain politics as well participation in social protests and movements; Touraine, 2011). People who participate in such movements and decision-making processes, despite their ignorance, are inclined to be guided by certain emotions or habits. The present research shows that there is a clear tendency for those with positive attitudes toward the technology to support (conditionally) CCUS technology without knowing anything about it. The same is true in metropolitan areas of the West Pomeranian region. However, the inhabitants of those cities are outside the areas of direct investments in CCUS technologies. In contrast, in places where CCUS investments are likely to occur, the strategy of respondents without knowledge of the technology is different: they suspend their judgments about these technologies, identified here as uncertainty about CCUS technology.

Then, socio-demographic factors emerge as significant, indicating the presence of social areas of uncertainty about CCUS technology. In these areas, it is not so much the potential for trust, but rather the presence of distrust toward the technology that emerges – as reflected in responses to questions indicating a tendency to protest such investments. In rural areas and small towns where such investments can be located, the scale of this uncertainty is greatest. The survey shows the potential existence of the NIMBY problem in medium towns. Still, no important statistical correlation exists between the NIMBY factors and the place of residence. By contrast, in rural areas and small towns, there is more uncertainty than an aversion to new technologies, implying a kind of silent NIMBY potential.

7 Benefits, Risks and Protest

Local Knowledge and New Energy Technologies

Introduction

This chapter focuses on risks and benefits, trust and protests. Those issues are well-recognized in public acceptance studies and here they are analyzed in specific social situations and concrete local communities. This issue is relevant to technology solutions such as UGS, but also to other technologies that require larger local investments (wind farms, photovoltaic farms, biomass plants).

Local contexts are always more complex than the national context. Acceptance of a particular technology at the level of national discourse may be only one of many and not a decisive factor in attitudes at the local level. This problem is well recognized in research addressing social protest, including NIMBY-type attitudes described in the first chapter. Therefore, the results of the qualitative research presented here should be viewed as a certain model for proceeding at the local level in diagnosing the determinants of perceptions of opportunities, barriers and acceptance in a given environment adopting investments in new energy technologies. Information on how specifically CCS and UGS are perceived in the surveyed communities is some illustration of the process of communication of the new technology and its reception by the respondents.

Ignorance and social acceptance of CCS and UGS in local environment

In the conditions of major ignorance on CCUS, social and cultural conditions, including local ones, are crucial for social acceptance. The comparison of different reactions to the experts' presentations during three deliberative meetings shows to some extent the mechanism of declaring the knowledge and ignorance of CCUS. The feeling of competence appeared among those who have moderate knowledge about energy technologies (without specialized solutions). People with higher or the lowest level of declared knowledge about the general technology declared ignorance of CCUS.

DOI: 10.4324/9781003607151-8

The qualitative data allow the reinterpretation of the results of quantitative survey in the direction that the actual knowledge on CCUS is lower in regional society than the one declared in the survey (19%). The only reference point of positive attitude for such technology is the normative trust in general technology, not traditionally understood knowledge as an element of social awareness of CCUS.

Already at this stage of the analysis it is possible to state that in Western Poland the project of raising the awareness cannot be realized because it should rather be a project of creating the awareness and then of trust building on these basis. It is illustrated by reaction of respondent on the information about CCUS:

> R2p: No, I'm shocked, that's the very first time, you know, I met the topic. I have to think it over. For sure I think, well, it should be explained in simple words, to everyone, socially, so the social resistance is not that strong, is it? [...] M: Do you think this has any realization potential? R2p: Well, I think, yes. It's only a matter of convincing people, and there's no problem with it. Everything is underground, I imagined it somehow different. I thought it was different.
>
> (Szczecin)

The statement above points out that even people who accept new technology take into account possible negative reactions of other people. In fact, the ignorance appeared together with the evident potential rejection of that technology because CCUS is very abstract and does not fit into everyday experience. The same attitude is shared by participants who are not open for new technologies at all:

> R5p: Such a vision, if someone said in the twenties, at the beginning of the twentieth century, that we would fly to the Moon, no one would believe it. Today, I am also skeptical about what you told us, what you presented. We talk about storing CO_2 and gas on the distance of 1.5 km, but nowadays we are unable to handle with methane in the mine, where from time to time, there are explosions and this methane is harmful – for the example the recent cases in Jastrzębska Spółka Węglowa.
>
> (Choszczno)

Only in Marianowo attitudes were more open on the CCUS technology among all categories of respondents, also the activists and local communities representatives not connected with the self-government. But these attitudes are a consequence of the fact that a natural gas storage facility had been planned 10 years prior to the AGaStor research. This local experience had familiarized the participants with this type of technology. At the same time it did not lower the doubts about finalization of another investment. It could be interpreted that the resistance to investment in UGS facilities did not result from a lack of trust in the technology itself, or ignorance. It resulted from previous experiences of unfinished investment.

Local contexts of social acceptance of CCS and UGS – needs and problems

The qualitative analysis of deliberative group interviews revealed that the situation of local communities can be analyzed within frames of needs, problems and dilemmas, and perspectives formulated by the participants.

Material values were indicated among the local community's basic needs such as: infrastructure development or number growth of investments and local jobs in the industry. From one perspective, the recognition of the commune's investment needs is a strong side of acceptance for potential investment in CCUS storage. From the other, the local community's expectations for such investment might be larger than it can bring. In this particular case, CCUS storage does not improve the local labor market situation.

Known problems that have been discussed in both places consider the choice between differently defined and contradictory needs of different residents' groups and between investment in the industry and preserving environmental and tourist values of the communes:

> R1p: We are the commune, we prepare different analyses, we prepare strategic documents, too, presenting the direction the commune should go. At the moment it is still developed, we rework it, we consult it, however what results from this analysis is that […] This development of tourism is one of the basics.
>
> (Marianowo)

> R7s: Considering tourist attractions, of course, Choszczno has a lake, there are few cool solutions implemented by the commune like building a bicycle trail around the lake. It surely helped in tourist matters.
>
> R2p: We don't have any industry, small entrepreneurship for sure, two larger workshops, we are from the commune with post-state farm areas. R3s: Besides, we talked before the meeting about green areas, golf courses, but we don't have workplaces here and we're a town that's very charming, but … R2p: With difficulties in development, we've tried with an industrial zone, but it hasn't worked out, unfortunately. M: Why? R2p: Generally, geography, the lack of larger roads nearby, we're close to state 10th (road), but that's all.
>
> (Choszczno)

The focus on the need for investment in industry and the feeling of being peripheral are crucial in the statement above. It may cause forced openness to acceptance for such technologies that can be used locally, under limited conditions of the commune. On the other hand – what further analysis revealed – there is a fear that such investment in CCUS might bring results quite far from the benefits from the industry expected by the residents. By now much more appreciated were "tangible" effects in the rise of the number of the jobs than the rise of income from taxes.

The perspective of building CO_2 storage seems to be neutral for these dilemmas in this sense that it will not change the state of both communes, even though it will cause the growth of their tax revenues. Such storage will not either decrease environmental values or increase the number of jobs. Probably, the storage will not be a factor repealing tourists – these areas are not particularly attractive tourist destinations.

Local communities' dilemmas influence the participants' perspectives on the future and their trials to specify development directions. The last ones look much better from the regional perspective. The respondents in Szczecin perceive their region as definitely pro-innovative while in local communities the future is treated schematically with the scheme's basis in tourism. In such a perspective of perceiving the future, investment and innovation are in the background.

Social awareness of CCUS technology: problems identified by DFGI participants

The participants' questions on potential investment connected with CCUS map the area of sensitive topics and significance in communicating solutions connected with the technology. In a sense they reveal these aspects of CCUS technology that may be most important in social perception. The analysis led to naming five of such areas: CO_2 transport from the capture point to the place of storing; the future (time of use) of CO_2 storage; the effectiveness of CCUS model, including designing the gas storage; the scale of usage of assumed type of injection; and the process of CO_2 injection. Therefore, it is crucial to inform about options of CO_2 transport, and to give the exact time horizon of the investment, including the state of the storage and the environment after ending the investment. It is worth remembering that some technical solutions, e.g., geological modelling of the gas storage, are far from common thinking and imaginations about it.

There are four thematic areas in analysis of the problems defined by the participants: problems with specificity of investment in CO_2 storage; problems with the realization of such investment; problems connected with the general political context of implementation of CCUS in Poland; problems with social reactions to this kind of investment.

It seems that the way of communicating technological aspects of CO_2 storage during DFGIs has worked out because most of the participants have been convinced about the safety of the technology. Nonetheless, in the context of CCUS, it is more as an imagined concern than measurable and controllable problem. Similarly, the presentation of CCS and UGS technologies has removed distress about eventual pollution that are part of potential nuisances of technological investments. The participants were convinced that the investment is neutral for the environment; however, they were aware of stereotypes of such investments as lowering tourist attractiveness of the surrounding. This problem is a direct consequence of the lack of information about CCUS technology in public discourse.

Another problem is connected with the fact that both in researched communities as well as in the whole region there are numerous initiatives for energy technology that are not finalized. Former experiences with unfinished offers and investment projects are the factor strengthening the participants' skepticism not for CCUS technology but for chances for realization of the investment.

The list of specific problems to solve in the context of localization of CO_2 storage is opened by the issue of transporting this gas. There has been strong negative reaction on road transport (with only few supporting opinions). The non-burdensome solutions (road transport means traffic jams, noise and destroyed roads) have been stressed during the discussion as the only acceptable for the participants. This allows us to assume that local communities will accept only the usage of pipelines.

Also, one-time use of CO_2 in the storage is problematic. CO_2 storage has been associated with a composition site for the substance. It limits the perspectives for possible benefits for the local community too. Contrary to gas depositories, focusing only on storing CO_2 eliminates the possibility of long-term use of stored substances. In consequence, this has led to the conclusion that one-time CO_2 injection causes that the solution is quickly insufficient to the problem of CO_2 emissions for entities storing it.

While the discussion on transporting allowed for a common search for the solutions acceptable for local communities, the one-time injection issue and the limits in receiving CO_2 (that will not be processed at this stage) has not led to any consensus. Concentration on CO_2 storing in presentation distracts from the core idea of the technology (the reduction of industrial emission of CO_2 into the atmosphere). More important is the question if such investment will be the source of sustainable income for the commune. One-time profit has caused concerns and associations with locating waste in the commune; however, such a term has not been used directly. The AGaStor project concept is based on the idea of a combined storage, where CO_2 will be used as a gas cushion for natural gas stored above. It should be considered if the information about CO_2 injection solutions could be broadened and information on CCUS technology could be presented as additional reinforcement of the investment attractiveness.

Moreover, the participants have issued other uncontrolled factors that have sources in social and political situations in the region and Poland. Recalling their own experiences, they have pointed three main threats to the final implementation of CCUS technology: law volatility, downturn in pro-ecologic technologies and political and economic uncertainty. They all seem to be connected, but each of them separately is a cause of formulating concerns in reality assessment in such investment. There are no juridical regulations on CCUS in Poland. The results of the research suggest that the appearance of such regulations will not cause a fast acceptance for the technology. Mistrust for the investment law and uncertainty about potential profits based on this mistrust should both be considered as key factors lowering the possibilities for building the trust for CCUS technology. Especially that the EU's decision on supporting this technology is also more a political project, without juridical solutions yet.

The list of predictable and at least partially controllable social problems consists of phenomena of local and regional social life. A specific challenge of promoting the investment in West Pomeranian Province is a major disappointment with investors. In all researched places, the information on CCUS technology has triggered memories of failed investments. Having this in mind, a potential investor should care not only for building the trust for the technology but also for the trust in themselves as a predictable business entity.

In the second place, there is a possible problem of associating CO_2 with waste. It results from the lack of knowledge about the context of CCUS technology; the connection between depositories and industrial CO_2 emissions; properties of CO_2; its meaning and presence in the planetary ecosystem. This thread should be exposed more in the description of natural processes of CO_2 crystallization.

Table 7.1 shows that technological and investment problems are much less of a risk factor for CCUS acceptance than social problems. The latter remain unsolvable regardless of the features of the CCUS facilities itself. At this point, it should be emphasized that the problems associated with the socio-political context in Poland are of local importance. They result from the specific profile of political actions and legal practices in this country. On the other hand, problems with the social response to CCUS can be regarded as a universal phenomenon. Lack of trust in investors is a manifestation of the low trust in business observed in most countries of the world. IPSOS Research in 2023 reports that for 31 countries in the world only 25% of people describe business leaders as trustworthy. Only in Indonesia, Thailand, India and Mexico the trust in business is over 33% (IPSOS, 2023, 29). The same situation is with the perception of CO_2 as a waste, which associated CO_2 storage with an environmental problem rather than a resource (Cynk, 2022). This way of thinking about CO_2 storage appears in DFGI local discussions and in regional interviews independently to the presence of the same argumentation presented in public discourse (see Chapter 4, 44).

Table 7.1 Character of local problems connected to investment in CCUS and UGS

	Specificity of CO_2 storage	*Realization of the investment*	*General socio-political context in Poland*	*Social reactions to this kind of investment*
Solved during DFGI discussion	1/ Safety of the UGS 2/ Nuisances of the investment			
Perceived as solved after DFGI discussion	3/ Losing touristic attractiveness of the region/ locality	1/ CO_2 transporting (road transport or pipe one) 2/ Short-time usage of CO_2 storage (and short time benefits)		
Unsolved			1/ Law volatility 2/ Downturn in pro-ecologic technologies 3/ Political and economic uncertainty	1/ Distrust to investors 2/ CO_2 treated as a waste

From NIMBY/WIMBY to OWMA in the qualitative local research

The NIMBY phenomenon is widely recognized in studies on social acceptance for both CCUS technology and generally in the context of local investments of industrial or any other public character that can be considered as potentially burdensome for the local community. M. Castells noticed (1997, 62) that such social movements combine conservative attitudes of local communities that strive to maintain the status quo. The WIMBY attitude, presented in the first chapter, stresses that the opposition is not the resistance toward new ideas, it is rather focusing on the privilege of local self-government communities to make decisions concerning the managed territorial units and its common interests.

The participants from Marianowo and Choszczno recalled examples of local protests that occurred in their communities. History of these protests was well known, and the participants reacted to them quite emotionally, which proved the real social meaning of such movements. It was visible in the description of protests against biogas plant, presented by participants from Marianowo:

> R3s: They wanted to produce fertilizing material, right? R1p: No! R7s: It's a biogas plant, well, components from corn! R2p: The main problem was the odor. R1p: It was the odor. That commune [L] R3s: The meetings were organized … R1p: Protested that hard. R3s: … the whole investment

> was supposed to be in the Marianowo commune, but next to the border with another commune. R2p: We didn't smell it that way [laughter]. R3s: We heard them downstairs, they shouted so loud, right? The meeting was organized, a lot of people came, but this company simply withdrew, right? There was so much public opposition.

The failed public consultation process mentioned above resulted from a combination of traditional NIMBY with the participation model used in European Union countries. In this particular case, it is worth stressing the motivation of protesters – it was a potential nuisance of the investment for people living in its surroundings, lowering the quality of life. This thread also appeared in further analysis. It can be interpreted as a kind of rational recognition of local interests typical of the WIMBY movement.

Two other examples of protests from Choszczno revealed further contexts appearing in local protest movements. The first one considered garbage incinerator:

> R4p: ... doesn't add up, we want to locate industry, but there will be none. There is no place for it and no social permission. M: No social permission? R4p: No permission, we can say straight, no permission. Of course, we do have problems connected with ... M: Why's that? R4p: It's historical, right? M: I understand that these are separate opinions among you? R4p: No, no, no. Generally I can say. There was an idea 20 years ago that considered incineration. Then there were protests, it even led to the situation that the mayor who had this idea, didn't win next elections. Right? He was unaccepted.

Besides the context of the above quotation, it revealed involuntary emphasis on the emotional meaning of protest movements. Events from almost 20 years ago were given as an example of social protest. The permanence of the memory of such practices also proves the real potential of protest movements in this commune, and perhaps also in the entire region.

Shifting to the level of everyday reactions of local communities revealed some of the mechanisms that may lead to protest movements' appearance. The first mechanism is **the role of individual interests.** Attitudes to new technology may have very individual sources, sometimes it is potential competition between municipalities. It was visible in the dynamics of earlier protests against wind farms:

> R2p: Maybe it didn't bother people that much, maybe there was also a little jealousy that neighbors leased land and they didn't, right? It can be an argument, too. Jealousy, because "I can't," "my land is useless and neighbor's isn't," so "the noise will disturb me," or anything else, right? It seems to me that these wind turbines were beneficial, because of the access roads to them, beautifully done field roads, we don't have such roads in our commune as in the next one. Because the investor straight

> away, really beautiful field roads are led to the wind turbines. As I say, the investor took care of it. We don't have such local roads in our commune as they have there.

Emotional reaction and very individual interests turn the attention on the very new version of the protest not described in the NIMBY or WIMBY model.

Besides major general reluctance to the investment, **the investors' attitude** was pointed as the second mechanism strengthening protest phenomenon:

> M: So you have experiences with different technologies? R4p: Well, yes, yes. Generally, the investor started from the wrong point, it seems. He didn't want to change anything later. As this case grew larger, it appeared that, eventually, that he didn't want to unwind it. And it died. But it looks like there were new technologies, cleaner technologies, better thought over, I think, this could be applicable here.

The investor's ability to be flexible, to adapt to the needs of residents, to take into account their concerns, to inform in a competent way, are all desirable features that are expected not only by the residents but also by local decision-makers. Local politicians, although they are aware of the limitations of the local community, are also aware of representing it and having the right to set specific requirements for the investor to have certain expectations. Those narrations come to the WIMBY nature of the protest when it stresses community needs (which could be understood as local interests). On the other hand, the need for detailed information for the residents of the community could be tied with the need for individual acceptance of proposed changes. This is why the participants themselves treated the deliberative group meetings in this particular way – as an expected form of communicating on CCUS technologies:

> R4p: I'm glad that this is substantive. Generally, in this Internet madness, all sorts of fake news, there's a lack of information. This is – I consider it in this way – the biggest advantage of our conversation.

The third mechanism is emotion management. This mechanism comes from the fact that each of the communities is a part of a wider socio-political environment from which the practices of building the concerns about new technologies emerge for political purposes:

> R5p: Our task, of those who decide somehow what will be done in our region, is to verify in the first place, not to scare [interrupted]. R4p: But this is the information, right? R5p: Not to scare with this biogas plant, not

> to scare with the wind turbines of course, that a PM or senator came and scared how those wind turbines were bad, how ugly, how noisy and how they hurt, cancers etc. People need to be told the truth. And people, of course not all of them, not 100%, but the majority.

This mechanism had – as it is visible – the source outside the local community and investor. It resulted from the fact that each of the communities is a part of a wider socio-political environment from which these mentioned practices emerge. And it focuses the question of local resistance in energy technologies from a new theoretical perspective besides NIMBY/WIMBY opposition.

Conclusions

It can be observed that the local problems which affect the reception of CCUS technology are determined primarily by the social and economic situation of specific communities. It changes when respondents consider CCS and UGS technology itself. Then they point to a broader list of problems related to technical solutions connected with CCUS, national laws and policies and problems that relate directly to social attitudes toward new technologies. The latter problems are distrust of business leaders and the treatment of CO_2 as waste. It is important that – according to the participants of DFGIs – these social problems are seen as particularly difficult to solve. At the same time, the data on social trust and environmental awareness indicate that these social problems are of more universal nature.

Qualitative research confirms the emergence of both traditional forms of protest attitudes as well as more moderate ones, described as WIMBY. What is particularly interesting and new in the research presented here is the emergence of new forms of protest as well. These forms of protest – described by the respondents – can be identified as belonging to a new understanding of civil society, as postulated by Alain Touraine (2011, 232) and represent emotional and ad hoc organized movements. Touraine indicates that the center of social life has shifted from thinking in terms of society toward the person as subject (Touraine, 2011, 160–2). The perception of social problems, including technological ones, is done not so much in terms of the common good but in terms of the freedom, reflexivity, relations and subjective rights of persons (Touraine, 2011, 182–8). All potential social conflicts are thus considered from the perspective of subjects in terms of violence – the imposition of restrictions on freedom, relationships and rights. "New new" social movements that emerge as a consequence of this social paradigm shift have other sources of protest. This attitude can be called "Only With My Acceptance" (OWMA). Those movements organize themselves in new forms characterized by ad hoc mobilization, lack of massiveness and the emotionality of the message (Touraine, 2011, 232).

But this attitude is still one among others. Residents, being aware of the complexity of decision-making processes but also of their participation in it, can easily adopt a WIMBY-type attitude. Previous studies show that it is more flexible than the classic NIMBY, but the results of our research also raise the question of whether it is not an initial and learned attitude in a society where there is widespread participation in decision-making processes.

Then NIMBY/WIMBY attitudes are not replaced by OWMA but coexist with each other, mixing – especially in the processes of argumentation, and the management of emotions and indifference. It is noticeable that representatives of the local government clearly emphasize that inadequate communication between the investor and the residents is a factor that causes a shift from indifference to resistance to investment. In such a situation, the local government is rather inclined to respect this lack of acceptance among residents. Here, the postmodern nature of the protest movement, which has a very strong emphasis on emotions and individual interest, may be a factor in determining concessions on the part of the authority.

The three main resistant social movements could be described by their individual and collective aspects (see Table 7.2).

The knowledge about such differentiation of the attitudes against energy technologies leads to questions on some communicational processes which could avoid the rejection of local investments in such technologies. The DFGI meetings in their last part were focused on the searching for some models or mechanisms for effective communication and trust-building. Description of such mechanisms and their effectiveness are presented in Chapter 8.

Table 7.2 NIMBY, WIMBY and OWMA attitudes

	NIMBY	*WIMBY*	*OWMA*
Individual	Fear of change	Ignorance the over-local needs	Emotional involvement of the subject
Collective	Collective identity resistant against global flows	Local interests strength by self-government rules	Resource for the subjects to organize a protest
Empirical manifestations	Preservation of status quo more valuable than future benefits	Balance of group gains and losses, recognition of ecological values and local needs	Recognition of individuals' opinions and feelings

8 Communication and Change

Analysis of the CCUS Promotion Meetings

Introduction

The first part of the chapter is based on qualitative research – DFGI (described in methodological annex in the introduction). After the recognition of main local problems and attitudes toward CCUS, the interviews focus on the way of building trust toward new technology. The aim of that part of the research was to search for possible and acceptable ways to communicate CCUS technology. Discussing with the local experts, we asked them about potential reactions of other inhabitants of local communities on the information about CCS and UGS investment presented during the DFGIs. The mechanism of building trust recognized in this procedure was compared with other concepts of communication of CCS (Otto and Gross, 2021). On this basis, the local meetings with residents of Suliszewo and Marianowo were held in April and May 2023. During the meetings with residents, the idea of the AGaStor project was presented, as well as information about the technology for capturing and injecting carbon dioxide into geological layers as a cushion for storing natural gas. After the presentation, experts answered any questions and concerns that arose. The evaluation surveys were conducted after those meetings. Their results are presented in the second and third part of the chapter devoted to changes in knowledge about CCS and UGS and reconfiguration of attitudes toward those technologies.

Communicating and trust-building mechanisms for CCUS

The last element of the analysis of DFGI data fulfills the recognition mechanism of building trust for CCUS as a way of communicating about this technology and about investments connected with it (CO_2 storage). We can distinguish four mechanisms of building the trust for CCS and UGS technology based on that qualitative data. As it was shown in the former chapter, the issue of safety

DOI: 10.4324/9781003607151-9

was very well perceived by the participants. In Marianowo, **communicating the safety** was the main trust-building mechanism for the CCUS technology:

> M: What from the expert's presentation has convinced you the most? R2p: Little harm to the environment, major threats weren't shown. Because for example, with wind turbines it's known that the noise and so on, right? And in this case none of such direct threats are visible. R5s: Exactly, there aren't any major threats. It's known that in every investment, in every endeavor something may go wrong, but in this case there's no such serious threat. R3s: The technology seems to be simple, not complicated, it gives such an impression. R5s: I agree with my colleague – it's not harmful and I think that it probably won't disturb tourists. As the professor said that he learnt about this storage he lives next to when he was preparing, I think it won't affect our environment. And this is really important because we have "Nature 2000" and people search for such places. R8s: I agree with ladies.

In the case above, the example of the UGS storage located in Podkarpackie Province was crucial in the communicating process. Key arguments were: (a) references to existing analogous depositories, and (b) giving examples from own experience. The first element seemed fruitful for the use in building trust in CCUS. The second one should be replaced with the study visits of potential stakeholders in existing gas depositories. Such actions seem inevitable in the perspective of lack of knowledge about CCUS technology.

In Marianowo, the participants were connecting the issue of the safety of the technology with the problem of potential nuisance of the storage. In this case, the way of informing during the debate about **the lack of nuisances** was also an efficient mechanism:

> M: Anything skipped or missed? Questions? R7s: I think that if such a thing happened, there wouldn't be problems that we have, there's no garbage from it, no side waste. Because of the fact that we have the lake and real estate, we have piles of garbage. It's impossible to control weekend tourists. In this case there wouldn't be any side problems. R2p: There won't be any smells, and now it's a nightmare, let's not deceive ourselves. Noise. M: Little noise. E: Little noise. R2p: How disturbing was transporting that silage, right? To Germany. Didn't matter holidays, not holidays. R7: But remember that [company's name] is the biggest tax payer in our commune. […] Luckily we don't have meat processing […] and this is only temporary, and it's the biggest tax payer in the commune. R2p: … anyway these odors do disturb [interrupted]. R7s: It's the biggest tax payer in the commune.

Of course, as in any such case, the issue of costs (nuisances) was connected with potential economic benefits, and it had been exercised in the commune.

This local experience in Marianowo was the capital of trust for new investments. In the mechanism of trust—the same as when failures with former investments lower the faith in realizing future ones—the existence of present investments strengthens the belief that next ones are possible. This includes, as shown above, investments in areas different from power engineering.

The building of trust by pointing out benefits was less effective. Probably, because local communities have slightly different imaginations of what can be a material benefit for them (certain jobs in the local market). Still, the representatives of self-government were certain that **communicating the benefits** was effective strategy for building the trust for the technology:

> R4p: Suliszewo is a pretty good example. Suliszewo wants gas, there are problems with it. I've been to gasworks many times on gasification. That's the problem in Suliszewo. When I look at this model and there's Suliszewo and they want this gas for example. And I see that if there's a well, it would speed it up. This storage would be driven and the possibility of selling it. At the moment the gasworks says: "No, it's not profitable for us at all!" Really, this network connection, generally. M: So something for something. [voices from the room – Residents' questions will be for sure – what will the village have from it?] R4p: No, doctor, it's the XXI century and people must know. Of course, they have awareness [interrupted]. R2p: But presenting, you must present benefits to the village [interrupted]. Maybe not for a particular resident, but [interrupted] R6p: For the whole village, right? R4p: It gives a kind of a premise, what is now impossible … M: Here is a problem, because the benefits for the commune are visible, because it's a tax. However, this would be tragic if there was gas storage in Suliszewo and they would have no gas. R4p: In such cases it won't be accepted at all. It's fundamental. M: They don't have gas? R4p: They don't.
>
> (Choszczno)

Like in Marianowo, this quotation shows that there was a discrepancy between expectations of quite specific material benefits and the investor's offer. It can be assumed that these profiled expectations will be different in different localizations.

The last mechanism of building the trust in communicating about CCUS technology seemed **transparent and competent transfer of knowledge**. This issue was especially appreciated by the participants in Choszczno:

> R4p: I'd tell you differently, I'd tell you this way. I think the strongest side of this meeting of two sides is that the AGH has substantive supervision. R2s: Exactly, that's positive. That's most important. R4p: In my opinion [interrupted]. R2s: Yeah, that's the advantage. R4p: What people are afraid of with for example these wind turbines [interrupted]. R2s: Yes, there wasn't, there wasn't. R4p: I think all this procedure is positive, because of the institution itself – you, professor, the doctors, that's substantive supervision, with exact

> knowledge and not like "I think so." This is a study on physical experiments, these wells, the depositories do function, right? It's not unknown, invisible, it is [interrupted]. R2s: Although these are scientific statements, the language is very accessible, and that's important. R4p: I think this is really positive for a discussion with society. The matter of communication and who does it. R2s: Yeah, exactly. R4p: As I listen to you, professor and doctor, he convinces me because he has the basis for knowledge [interrupted]. Voices from the room: "Yes." R4p: The knowledge is a fundamental thing and what was missed [indistinct] is exactly something like this. Who creates the communication. Another thing is, of course, we don't have this knowledge, yes, because, you geologists and so on and so on. We lack these details that we don't know. It's a matter of forwarding this knowledge to someone…

At the beginning of this part of discussion, the authority of the university was mentioned but this could be generalized as presence of experts and their involvement in explanations:

> R2s: In my opinion this meeting, though online, was really effective. So repeating such meetings for more people, I don't know, is the cinema. [voice from the room – exactly].

The accessibility of the message was also appreciated in Marianowo (in this case it was based on visual materials). It appears that combining presentation, experts' presence at a place and online was supporting the effectiveness of communication on technology, including CCUS technology. However, the crucial conclusion for recognition of mechanisms of trust building was a combination of competences and simplicity of the message. It is fundamental for communicating on technology, and main tools for it are: (a) transparency in detailed information on CCUS technology and the usage of deliberative techniques, (b) referring to experts who can be identified as independent (e.g., from the investor) specialists in the technology.

Distinguished trust-building mechanisms are useful for organizing promotional meetings about CCUS technology. They can also be compared with Otto and Gross' (2021) recommendations for CCS communication (see Chapter 1). During our research we are concerned with "content and communication style" and the results of the study show that the crucial factor of efficient communication is "storytelling with links to everyday life experiences". This approach has proven successful in communicating security content of the CCUS project. The DFGI experience stresses the importance of such approaches as "transparent and egalitarian communication" and "communicating opportunities and challenges" which are in Otto and Gross' propositions. But according to our evidence, those approaches are in some opposition to the next two guidelines: "establishing positive attitudes toward CCS" and "avoiding over-consulting the community". This opposition does not apply to all aspects of communicating CCS technology; however, it does apply to benefits and risks. Even though these two aspects are usually combined in the literature, our respondents treated them clearly separately. Regardless of objective technological dangers and advantages, benefits and risks are socially constructed (Beck, 2012). Therefore, the way they are communicated is particularly sensitive to social conditions. In our perspective, establishing "positive" attitudes is rather a possible result, and not a communication approach. The recipients expect a neutral stance toward the communicated technology as a core of "competent knowledge transfer". The same conclusion may apply to the expectation of continuous and detailed information about benefits and risks. Also, equal treatment of benefits and risks may be positively perceived by recipients, although we fully agree that this means that they are always "carefully communicated".

How deliberative meetings change the perception of CCS and UGS?

The promotional meetings attended by residents of Suliszewo and Marianowo were the last part of AGaStor research. After the presentation of ideas and practice of CCS and UGS technology and discussion of the questions which appeared in the audience, participants were asked to fill out a paper evaluation questionnaire. The set of questions was similar to survey questionnaires realized in representative samples in the first stage of research (described in Chapter 6). A total of 31 evaluation questionnaires were collected. The data collected from the evaluation surveys were juxtaposed with the results of the survey research, which made it possible to estimate the directions of attitude change associated with the provision of information according to the trust-building mechanisms presented.

If we compare the results of the evaluation questionnaires with the quantitative data carried out in the West Pomeranian Province, it can be estimated that the information presented at promotional meetings improves the knowledge of participants about CCS and UGS technology (Figure 8.1). The small change concerned CO_2 capture, where association with the emission of this gas from industrial plants appeared in only 48% of respondents. The association of underground gas storage with being located under a layer of impermeable rock should be considered a better result. This association was noticeably weaker in the population as a whole (29%), while in the evaluation survey it was 55%.

However, as the achieved indicators are very low, only about half of the respondents declare correct associations. It was necessary to take into account the socio-demographic data of the respondents as a potential source of different reactions to the presented promotional materials. In the case of recognition of gas storage facilities as being located in deep geological strata, clear differences emerged between men (33% associations) and women (61%). However, apart from that, gender did not appear as a significant feature differentiating the knowledge and opinion of the respondents.

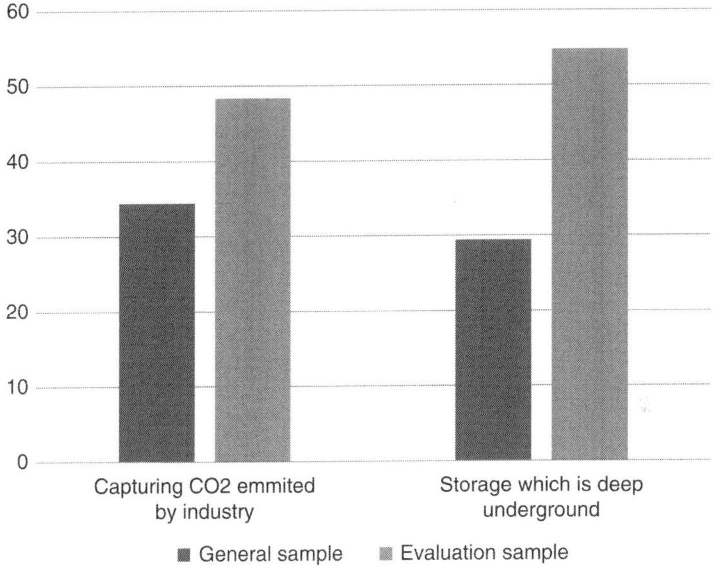

Figure 8.1 Comparison of knowledge about CCS and UGS between regional survey and evaluation survey (in %) [AGaStor Research]

In contrast, clear differences emerged in relation to education. The percentage of people with a university degree associating gas storage facilities with their location in deep geological strata was as high as 77%, with a vocational degree only 43% and with a secondary education 37% (this is not a linear correlation).

The importance of education was revealed in other data, including the fact that only those with a vocational education said they had not encountered the term CO_2 capture (28%) – just after a one-hour meeting devoted to this technology. These results suggest that the communicational mechanisms of building trust may have been less well suited to those with lower education and better suited to those with higher education. Remaining with the analysis of general knowledge of CCUS, it can be seen that the respondents rated the safety of this technology very highly (97%).

However, overall, a more detailed analysis shows a significant increase in knowledge about CCS and UGS. A comparison of this knowledge was made by comparing the results from the evaluation survey with those of the 25-person group of respondents in the regional survey, whose demographic and social characteristics exactly matched those of participants in the evaluation survey (Table 8.1). Responses stating that CCS is part of the European Green Deal were 50% higher than the regional group (adding "definitely yes" and

Table 8.1 Perception of CCUS as green technology in regional survey and in evaluation after promotional meetings (in %)

	General sample				*Evaluation sample*			
	Definitely no	*Rather no*	*Rather yes*	*Definitely yes*	*Definitely no*	*Rather no*	*Rather yes*	*Definitely yes*
CO_2 capture and storage technology is part of the European Green Deal	60.9	8.7	30.4	0.0%	4.8	4.8	42.9	47.6
The technology of capturing and storing CO_2 is one of the technologies of obtaining clean energy sources	56.5	21.7	17.4	4.3	11.1	11.1	44.4	33.3
Underground CO_2 storage is a safe gas storage technology that can be used	52.4	23.8	23.8	0.0	0.0	18.8	43.8	37.5
Underground CO_2 storage is an effective technology to reduce the presence of this gas in the air	47.8	21.7	26.1	4.3	11.8	5.9	35.3	47.1

Source: AGaStor research.

"rather yes"). Indications of CCS's connection to clean energy increased by 40%, and indications of the technology's safety were indicated by 43% more participants of deliberative meetings. Finally, 33% more respondents after the deliberative meeting recognized that this is a technology that effectively reduces CO_2. It can be seen that the increase in knowledge was primarily about the environmental aspect of CCS technology. It is worth remembering that in the regional survey this type of association was very weak. In fact, it can be said that the lack of belief in the ecological nature of CCUS may be the weakest element in the communication of this technology.

At the same time, the direction of this change in social awareness of CCUS also relates to perceptions of benefits and risks. The change in attitudes toward risk is noticeable both in the increase in the belief that CCUS is a safe technology and in the essentially universal belief of participants in deliberative meetings about this safety (what was indicated by the data cited above). Among meeting participants, indications that CCUS brings more risks than benefits drop by 16%, while the belief that such an investment can benefit the community increases by the same amount. The belief in economic value is also seen in a 54% increase in support for the thesis that CCUS has environmental and economic benefits.

The change in awareness of CCUS technology, however, leads us to the question of its implications for social acceptance. Indeed, the trust-building mechanisms used were aimed at changing knowledge as well as changing attitudes.

Building trust as the transformation of attitudes toward CCS and UGS

The results of the analysis of social acceptance of CCUS technology show that familiarization with that technology and enlargement of individual awareness about it results in a more cautious attitude toward the negative consequences of locating CO_2 storage facilities. In evaluation research, respondents who strongly focused on pro-technological values are unlikely to want such storage to be located in their own community. In the region-wide survey, the respondents oriented on such pro-technological attitudes had a more positive attitude toward nuisance connected with new energy technology and accepted CCUS investments in their surroundings (see Chapter 6). It should therefore be recognized that deliberative meetings where the mechanisms of building trust are used reduce not only negative stereotypes. The positive attitudes that are not linked to knowledge of the technology but only on pro-technological values also disappeared in the evaluation surveys. In contrast, consideration of UGS investments in the context of group economic and energy benefits (at the local level) met with exceptionally high approval among respondents (73% of the total group).

Questions about support for CO_2 storage facilities used in the questionnaire also yielded higher rates of acceptance of this type of investment, both

Table 8.2 Acceptance of underground CO_2 storage vs pro-technological attitudes. Evaluation survey 2023 (in %)

Agreement on underground CO_2 storage facilities:	*Life quality: technological development (as the source of life quality)*				
I strongly support all such investments	*to a very low degree*	*Moderately*	*to a high degree*	*to a very high degree*	*Total*
Hard to say, I don't know	100.0	27.3	36.4		30.8
Definitely no		9.1			3.8
Rather no			9.1		3.8
Rather yes		36.4	18.2	100.0	34.6
Definitely yes		27.3	36.4		26.9

Source: AGaStor research.

in the group of those with pro-technological values and in the evaluated group as a whole. It is worth noting that, at this overall level, as many as 61.5% strongly support UGS-type investments with CO_2. This shift comes at the expense of those who in the regional surveys chose the answer "difficult to say", which was indicated as a sense of uncertainty about the new technology.

Table 8.2 shows that deliberative meetings have the potential to offset feelings of uncertainty. This does not always involve acceptance of certain disadvantages. It can be assumed that it is a greater awareness of the technology in question and a general acceptance of it, rather than an acceptance of all aspects of it. This leveling out of insecurity is also manifested in the rejection of radical attitudes of protest. Again, this change applies not only to protest attitudes but to the entire group surveyed.

Table 8.3 shows that only individuals are willing to adopt an attitude of generalized rejection of UGS technology. The uncertainty index applies to

Table 8.3 Rejection of CO_2 underground storage technologies vs pro-technological attitudes. Evaluation survey 2023 (in %)

Agreement on underground CO_2 storage facilities:	*Life quality: technological development (as the source of life quality)*				
I believe that any such technology is superfluous and should not be invested in	*to a very low degree*	*Moderately*	*to a high degree*	*to a very high degree*	*Total*
Hard to say, I don't know	100.0	20.0	27.3		24.0
Definitely no		40.0	27.3		28.0
Rather no		20.0	45.5	100.0	40.0
Rather yes		20.0			8.0

Source: AGaStor research.

almost ¼ of the respondents, while rejection of this type of protest is declared by 68% of the respondents.

Milder forms of rejection of CO_2 storage technology already appeared in the category of respondents with vocational education. All of them agreed with the view that this technology carried more risks than benefits (answer: definitely yes). This view was shared by single people with a university education (rather yes: 15% of the category) but almost half of those respondents chose the "hard to say" option (46%).

Strong and mild rejection of CCS and UGS technology might be compared with the two types of attitudes against UGS investments in local communities. There are traditional NIMBY and more sophisticated WIMBY attitudes mentioned in Chapter 1. Using the questionnaire from the regional survey and evaluation research, the NIMBY and WIMBY scale was built. The indicators for Not In My BackYard are answers on two questions:

"the stronger one": "Under no circumstances would I want such technologies in my county, even if they are safe" which is a definitive rejection of the CCUS.

"the weaker one": "In general, I consider it right to develop and use such technologies, but they should not be located in my county" the rejection of understanding the usage of the CCUS but still is against the technology.

Why In My BackYard attitude is indicated by the answers on other two questions:

"the stronger one": "I would agree to them if they bring material benefits" which is an agreement with benefits

"the weaker one": "I could agree with them, because they are "green" technologies that protect the climate" which agrees with the altruistic approach.

Worth to note, that NIMBY statements are negative in their construction (express disagreement) and WIMBY statements are positive in their construction (express agreement).

All these are on the scale from "definitely no – 1" to "definitely yes – 4" with "hard to say, I don't know" coded as 0. Therefore, if they are added, the higher the result is, the stronger the agreement with a given statement. So we built these two indicators by adding answers to these statements within signaled configuration:

Rate 0 mean the uncertainty and/or lack of opinion;

Rates 1–2 inform about rejection of NIMBY/WIMBY attitudes. That could be interpreted as a kind of positive attitude and social acceptance of CCUS technology;

Rates 3–4 describe ambivalence of the respondent's opinion about CCSU;

Rates 5–6 mean that respondent rather agree with both statements connected with NIMBY/WIMBY attitude – it is soft form of those attitudes;

Rates 7–8 describe strong form of NIMBY/WIMBY attitudes – respondent rather agree with one statement and strongly agree with another one (7) or strongly agree with both statements (8).

NIMBY values indicating rejection of CCUS numerically increased compared to the group in the regional study (Figure 8.2). However, it should be taken into account that in the regional survey as many as 72% declared ignorance and uncertainty (rate 0). The rate of uncertainty in the group with deliberative meetings is 50% lower. Only a minority of respondents who decided to have concrete opinions about CCUS adopted a position of rejection of that technology – a total of 27% among meeting participants have a soft NIMBY attitude, compared to 8% in the general survey. This allows us to assume that almost half of those people who potentially had no opinion before the meeting do not agree with NIMBY-type statements or treat them ambivalently (rate 1–4). In this sense, it can be considered that the communication technique used largely transforms uncertainty into non-acceptance of traditional forms of technology rejection described as NIMBY attitude.

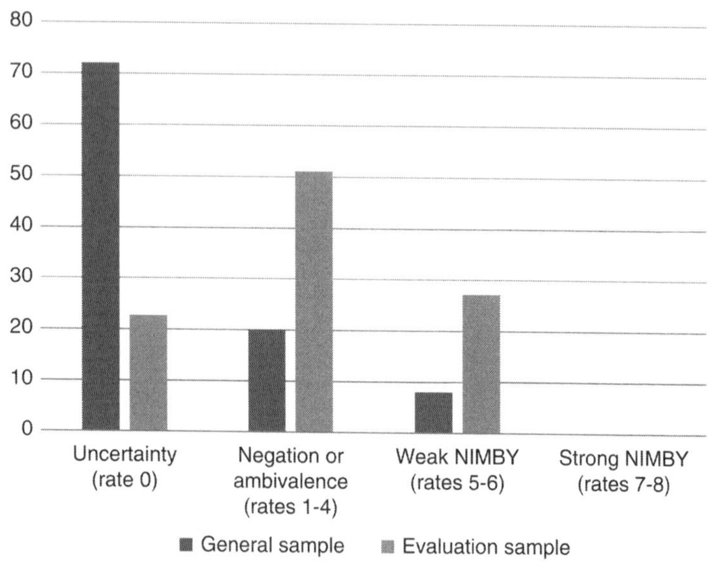

Figure 8.2 NIMBY attitudes – comparison regional sample and evaluation of deliberative meetings [AGaStor Research]

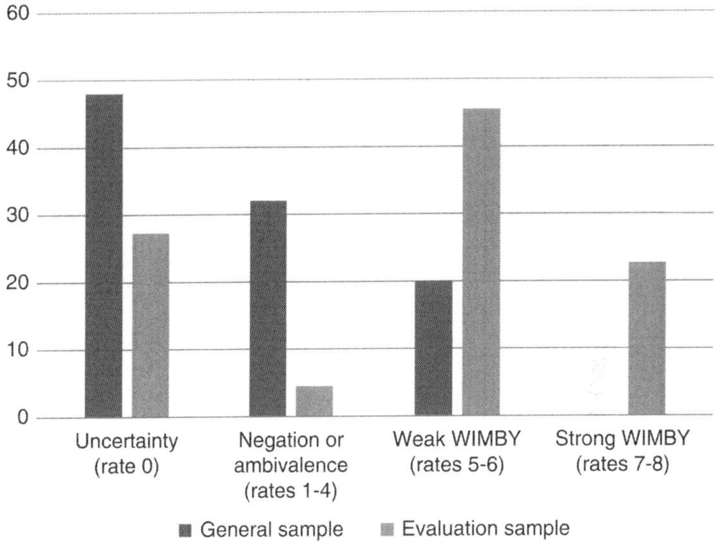

Figure 8.3 WIMBY attitudes – comparison regional sample and evaluation of deliberative meetings [AGaStor Research]

The WIMBY attitude, however, appears in the research in a different way (Figure 8.3). Here, the sense of uncertainty also decreases, but it feeds primarily into an attitude of conditional acceptance of technology, recognizing the priority of community benefits which is the content of WIMBY attitude. This is true for as many as 2/3 of meeting participants. Also important is the emergence of a clear category of people strongly stating the necessity of both economic and environmental conditions to accept CCS and UGS – that is, a strong WIMBY attitude, absent in the conditions of the regional sample.

Conclusions

The evaluation research shows that the deliberative meetings give the participants a basic knowledge about CCS and UGS as the new technology. Those meetings also increase the sense of security regarding technologies such as underground storage of natural gas or CO_2 in the geological layers.

At the same time, the research points out that there is a problem with tailoring the message about CCS and UGS to different groups of respondents. Information that is well absorbed by those with higher and secondary education seems inadequate to meet the cognitive needs of those with less education.

It should be noted that the provision of knowledge about specific technologies also translates into raising the respondents' requirements for the quality of investments in CCS and UGS. Considering the increase in public participation in investment processes, this can be considered a positive result of the popularization of technologies.

The deliberative meetings and trust-building mechanisms primarily reduced the level of uncertainty and let the participants have an opinion about CCUS. The effect of that change is the softening of radical, traditional rejection toward CCS and UGS, which are expressed in the NIMBY attitude. But, on the other hand, the rising of social awareness triggers mild forms of rejection – WIMBY attitudes.

Therefore, the communication practices and trust-building mechanisms used in deliberative meetings should be regarded as an element that accomplishes just such tasks: it provides knowledge about CCUS and mutes direct negative reactions against the technology as such. Social acceptance at the level of recognition of the value of local investment in CCUS remains an unresolved issue. However, this issue goes beyond the characteristics of CCS and UGS technology itself. Rather, it is a matter of recognizing to what extent the economic values generated by such an investment meet the real needs and expectations of the local community. The fact that in the communities studied these needs diverge from the opportunities associated with CCUS was already indicated in Chapter 7. This is a local condition, however, bearing in mind that CO_2 storage is an investment that is not only minimally invasive to the surface natural and social environment but also does not offer, for example, a large number of jobs. This may be a problem not only in Western Poland.

Conclusions

To conclude this book, we can start with a provocative question: why is there all this talk about social acceptance of CCUS at all? Well, to clearly emphasize the fact, the problem of social acceptance of any technology related to energy transition is inextricably linked to the issue of civil society and the democratic participation of citizens in economic processes. The participatory dilemma, mentioned in Chapter 1, is that the energy transition is for the people, but the inclusion of local communities in decision-making processes significantly delays the adoption of effective technological solutions. We believe that this dilemma under the conditions of the European Union should be resolved in favor of local communities. This conviction was the source of adopting a particular cognitive and methodological perspective.

Deliberative methods "allow for a more democratic process of knowledge development through the voices and experiences of local people. Pluralism of opinions can also be argued to be essential for democratic processes, as, through the expression of diverse voices and social interactions, societal conflicts can become more apparent and accessible to everyone" (Nielsen et al. 2022, 15). In a broader cognitive context, this approach to the voices of communities and individuals means incorporating their knowledge into scientific discourse (Geertz 1993). Disseminating research findings also gives these local voices a chance to participate in public discourse.

The presented case of AGaStor project in West Pomeranian Province in Poland is an interesting study of a society that has not yet engaged in a collective "symbolic coping" with technology that – due to the specific nature of the country's energy structure – is inevitable within the next decade. But in the Polish public discourse, this topic does not exist: include both official strategic documents of the Polish government, the level of interest of traditional mass media and social media. However, it should be expected that the launch of any CCS or CCUS project will force the need to work through this issue within the public discourse, at least on a local level. West Pomeranian Province is a more "westernized" region than other regions in Poland. It could be observed in social-economic data and in the fact that regional societies accept pro-environmental values. In the regional economy, the proximity to

DOI: 10.4324/9781003607151-10

projects related to the energy transition, namely numerous wind farms that have sprung up over the past years.

But, on the other hand, the local determinants of Choszczno (Choszczno county) and Marianowo (Stargard county) diverge from the regional characteristic. Statistical characteristics present Choszczno as a demographically and economically weak area, with unused labor potential, lower social and economic activity compared to the region and compared to Stargard. It seems that these social features may be a factor making social acceptance of CCUS technology easier when it will be presented as pro-development investments. Marianowo, on the other hand, was selected in 2009 as the location of the LPG warehouse and though this plan was abandoned after a few years, the positive attitude of local authorities to this idea was noticeable. In both counties local communities are more right-wing oriented in people's political choices, in the sense of skepticism toward the free market, orientation toward religious values, traditionalism and rurality as well as eager lifestyle change to a more ecological one.

One should be aware that locating this type of investment in small, weak local communities that are experiencing negative demographic phenomena is not and will not be a Polish specificity, but at least European one. It does not change the assumption that democratic procedures for the development of the energy transition should be considered as important as the issues of energy security and the Net Zero Carbon Emissions. The political turbulence surrounding CCUS, but also other energy solutions (such as nuclear power), shows that miscommunication of the technology and lack of public support can be a serious source of inhibiting economic and environmental goals.

The issue of CCUS is a huge challenge facing the European Union. Although it is only one of many technologies related to the energy transition, it is treated as a key CO_2 mitigation solution to facilitate the transition process and achieve Net Zero Carbon Emissions. This is because of the demand of EU countries for immediate carbon reduction in the face of lengthy and complicated processes to achieve full energy transition. The current dependence on fossil fuels makes EU countries unable to abandon carbon storage technology, while facing the fact that it is a solution unrecognized by the majority of public opinion in these countries. The issue of CCUS is also a matter of energy security. Viewed in the context of the energy and economic systems of the eastern part of the European Union, the technology seems particularly beneficial and necessary for ensuring energy security in central Europe and the Baltic Sea basin.

In 2030, more than half of the energy produced in Poland will still come from fossil fuels. The diversification of sources of these fuels, accelerated primarily by the war in Ukraine and the departure of European countries from coal and gas from Russia, translates into the growing strategic importance of hubs such as the LNG Terminal in Świnoujście, but also the infrastructure for transmitting and storing imported fuel. Solutions aimed at reducing the cost of technologies used in this area – even if these are strategies that mitigate the transformation ultimately leading to a transition to renewable energy sources

Printed in the United States
by Baker & Taylor Publisher Services

in the long term – are becoming particularly important. In the case of the AGaStor project assumptions, the expected effect has two consequences: (a) the use of captured CO_2 as a cushion reducing the costs of recovering stored natural gas, and (b) the use of captured greenhouse gas for this purpose reduces the emission intensity of the entire national system.

It should be emphasized, however, that geological conditions, the geopolitical situation, global logistics chains, the long-term state strategy, as well as the historical conditions of the energy production system in Poland translate into locating preferably this type of investment in the region, which is the national leader in energy production from renewable sources, has significant protected natural areas and derives a large part of its income from tourism. What is more, the communes located on a geological structure that is strategically important for storing natural gas (on the one hand, favorable rock formations, on the other, proximity to existing infrastructure) are struggling with economic and social problems, which seems to be typical for the eldering societies in the EU. For them, this theoretical investment is a potential opportunity to empower local needs in the process of negotiating the terms of social acceptance or the lack of a potential protest.

In conclusion, we consider what lessons can be learned from local research in northwestern Poland for the general understanding of the meaning of CCUS and the social acceptance of energy transition in contemporary Europe and the world.

In the chapters devoted to issues of a theoretical nature, we have tried to bring the reader up to date with the latest knowledge in the field of social acceptance of new technologies, energy security and policies related to carbon capture and storage. We are confident that against this theoretical background we have succeeded in the remainder of the book, based on empirical research and specific carefully analyzed examples, to verify the existing knowledge regarding public acceptance of CCUS technology deployment.

First, these analyses show that CCUS technology is specific in the context of other energy transition technologies. Its specificity lies in the fact that it is a socially unrecognized and abstract technology from the perspective of everyday experience. Thus, addressing the social acceptance of CCUS is a kind of laboratory for recognizing how European societies react to new energy-related technological solutions that are complex and difficult to accept.

Another important result of the research presented in the book is to note the fact that at the local level, some factors of social acceptance are specific to a given community or country, while others have the character of universal expectations and needs toward investments in new energy technologies. Noting these needs of a universal nature makes the research in local Polish communities relate to the broader context of energy transition in the modern world and democratic highly developed societies.

Undoubtedly, such a broader context, noted at the regional as well as local level, is the problem of uncertainty. This phenomenon relates to Beck's well-known theoretical concept of risk society. Our research shows in detail in which places of social reality and under what social conditions a sense of uncertainty develops. First of all, it arises due to the lack of access to detailed knowledge about new technologies in the energy industry but also to general knowledge about the energy transition.

One of the key challenges of implementing the energy transition in European societies is attitudes of social acceptance of new technologies – the central issue of this book. Our research confirms that there is a very complex web of social attitudes in the modern world. These include both the propensity to adopt certain technological solutions and the complex phenomena of resistance to these technologies. Our research shows the coexistence of many types of attitudes opposing new technologies in the energy industry: from classic NIMBY-type attitudes, through – already recognized in studies in Scandinavian countries – WIMBY attitudes, to completely new emerging phenomena, described by us as OWMA.

The success of the research presented here is the recognition and more complete description of these three types of attitudes against new technologies. We were also able to show the temporal connections and worldview distinctions between these attitudes.

The research also included planning such ways of communicating new technologies that would address public concerns. In this regard, the literature suggests a number of solutions for the correct communication of CCS technologies. Conducting specific research in local societies made it possible to see which of these solutions are effective and which should be modified. Detailed evaluative research has established that an increase in knowledge toward a particular technology causes a significant change in attitude. Without communication action the CCUS technology is not recognized as connected with the CO_2 emission mitigation. But we can see some changes in the RET perception because 10 years ago Brunsting et al. (2013) wrote:

> Agreeing that climate change happens and that measures should be taken does not imply agreement on CCUS as a suitable method to curb climate change. (…) If CCUS is used it should be a short-time solution implemented along with an exit strategy so as to not divert attention from other options which are perceived to be more sustainable in the long-term such as renewable energy. (…) They could also think of international downsides such as becoming a garbage dump for European CO_2 emissions. To the Polish group, therefore, one of the conditions for accepting a local CCUS project was that only CO_2 produced in the region would be stored. (…) In Poland, location of the storage site raises concerns with the participants about possible loss of the value of surrounding real estate. (…) It should be made clear with whom the responsibility for the project lies. The Polish participants mentioned that the government should financially support the development of CCUS and generally should provide clear legislation on CCS.

If experts and scientists do not question the issue of climate change, such a problem is still valid for public opinion. It is another plane of discussion about empowerment and emancipation of local communities – the problem of possible change of lifestyles and a question on the decision-making process. In this perspective, the perception of CCUS technology is more a tool for conducting arguments than an end value (goal) in itself.

This change, however, does not mean acceptance of CCUS technology but a shift toward more benign forms of resistance – from NIMBY to WIMBY. This direction of change shows that promotional and informational campaigns (even with the deliberative component) are not a sufficient way to build public understanding of the energy transition. The transition argumentation effectively reaches the respondents, but locally, specific needs and group interests are formulated beyond that. The location of investments in CCUS, and especially UGS facilities, from the perspective of peripheral local communities, can be associated as internal colonization that violates "feelings of social justice" (Wagner and Götz, 2021, 12). The disparate forms of attitudes against new technologies – NIMBY, WIMBY and OWMA – show that even within a single community, people can appeal to different notions of what social justice is.

The research presented in this book, therefore, highlights two pressing issues related to CCUS and the social acceptance of new energy technologies in general. The first is the need to develop sustainable institutional arrangements for continuous and clear communication of the energy transition process. It seems that this cannot be at this stage a spatially and thematically decentralized process of communicating its various aspects and different technologies. In this communication, it may be important to clearly indicate the links between technological changes, lifestyle changes and social safety.

The second challenge that should be subjected to constant and extensive research is the diversification of attitudes against new energy technologies. It seems that past research, and in part our practice in the AGaStor project, has been oriented toward improving the way new technologies are communicated. The transformation of NIMBY-type attitudes into WIMBY-type shows that this way of thinking is insufficient. Research on the rejection of new technologies should develop in the direction of establishing socially constructed rules of justice. It seems the perspective which takes into account the categories of social justice and solidarity is a challenge for future research on the social acceptance of new energy technologies.

Literature

Arias A., Feijoo G., & Moreira M.T. 2023. Advancing the European energy transition based on environmental, economic and social justice. *Sustainable Production and Consumption*, 43: 77–93. https://doi.org/10.1016/j.spc.2023.10.013

Ashworth, P., & Quezada, G. 2011. Who's talking CCS? *Energy Procedia*, 4: 6194–6201.

Babbie, E. 2013. *Podstawy badań społecznych*. Warszawa:Wydawnictwo Naukowe PWN.

Batel, S. 2020. Research on the social acceptance of renewable energy technologies: Past, present and future. *Energy Research & Social Science*, 68: 101544.

Batel, S., Devine-Wright, P., & Tangeland, T. 2013. Social acceptance of low carbon energy and associated infrastructures: A critical discussion. *Energy Policy*, 58: 1–5.

Baxter, J. 2020. Health and environmental risk. In: Kobayashi A. (ed.), *International Encyclopedia of Human Geography* (2nd edition). Amsterdam: Elsevier, pp. 303–307

Baxter, U. 2012. *Społeczeństwo światowego ryzyka: W poszukiwaniu utraconego bezpieczeństwa*, Warszawa: Wydawnictwo SCHOLAR.

Blumer, H. 1936. Social attitudes and nonsymbolic interaction. *The Journal of Educational Sociology* 9(9): 515–523.

Blumer, H. 2007. *Interakcjonizm symboliczny*. Kraków: NOMOS.

Bradbury, J., Ray I., Peterson T., Wade S., Wong-Parodi G., & Feldpausch A. 2009. The role of social factors in shaping public perceptions of CCS: Results of multi-state focus group interviews in the U.S. *Energy Procedia*, 1: 4665–4672.

Bralczyk, J., 2007. *O języku propagandy politycznej lat siedemdziesiątych*. Acta Universitatis Upsaliensis.

Brown Weiss, E. 2011. The evolution of International Environmental Law. *Japanese Yearbook of International Law* 54: 1–27. https://scholarship.law.georgetown.edu/facpub

Brunsting, S., Pol, M., Mastop, J., Kaiser, M., Zimmer, R., Shackley, S., Mabon, L., Howell, R., Hepplewhite, F., Loveridge, R., Mazurowski, M., & Rybicki, C. 2013. Social site characterisation for CO_2 storage operations to inform public engagement in Poland and Scotland. *Energy Procedia*, 37: 7327–7336.

Celia, M.A., Peters, C.A., & Bachu, S. 2002. *Geologic Storage of CO_2: Leakage Pathways and Environmental Risks*. Princeton, NJ: Princeton University.

Charveriat, C. 2023. The Green Deal: Origins and Evolution, Geopolitical Study Group, after COP 27: Geopolitics of the Green Deal. https://geopolitique.eu/en/articles/the-green-deal-origins-and-evolution/

Cherp, A., & Jewell J. 2014. The concept of energy security: Beyond the four As. *Energy Policy*, 75: 415–421.

Child, M., Koskinen, O., Linnanen, L., & Breyer, C. 2018. Sustainability guardrails for energy scenarios of the global energy transition. *Renewable and Sustainable Energy Reviews*, 91: 321–334. https://doi.org/10.1016/j.rser.2018.03.079

Charmaz, K. 2003. *Constructing Grounded Theory. A Practical Guide Through Qualitative Analysis*. London: Sage.

Clulow, Z., & Reiner, D.M. 2022. Democracy, economic development and low-carbon energy: When and why does democratization promote energy transition? *Sustainability*, 14: 13213. https://doi.org/10.3390/su142013213

Cynk, K. 2022. *Znaczenie środkowoeuropejskiego dyskursu akademickiego i medialnego w rozwoju socjologii środowiska*. Rzeszów: Wydawnictwo Uniwersytetu Rzeszowskiego.

Dear, M. 1992. Understanding and overcoming the NIMBY syndrome. *Journal of the American Planning Association* 58(3): 288–300.

The Declaration of Energy Ministers. The Baltic Sea Energy Security Summit, Marienborg, Copenhagen, 30 August 2022.

Denzin, N. 1979. *The Research Act*. New York, NY: Routledge.

Devine-Wright, P. 2009. Rethinking Nimbyism: The role of place attachment and place identity in explaining place protective action. *Journal of Community &Applied Social Psychology*, 19: 426–441.

Devine-Wright, P., & Wiersma, B. 2020. Understanding community acceptance of a potential offshore wind energy project in different locations: An island-based analysis of ‘place-technology fit’. *Energy Policy*, 137. https://doi.org/10.1016/j.enpol.2019.111086

Dütschke, E., Wesche, J., Oltra, Ch., & Prades, A. 2019. Strategy CCUS. A viable solution for a sustainable future. Stakeholders mapping report. https://strategyccus.brgm.fr/sites/default/files/D3.1_STRATEGY%20CCUS_2019_08_28_Stakeholder MappingReport.pdf

Duszczyk, M. 2009. *Powstanie pierwszy komercyjny magazyn gazu w Polsce?* www.forsal.pl

Dziekoński, O.R. 2020. *Samorząd województwa na rzecz transformacji energetycznej dla klimatu i rozwoju regionalnego.* Energetyka Rozproszona, Zeszyt 3; https://www.energetyka-rozproszona.pl/czasopismo/energetyka-rozproszona-zeszyt-3/

European Commission. 2019. *Directive (EU) 2019/944 of the European Parliament and of the Council of 5 June 2019 on common rules for the internal market for electricity and amending Directive 2012/27/EU Brussels*, 2019, Brussels: European Commission

European Council. 2012. Fit for 55. https://www.consilium.europa.eu/en/policies/fit-for-55/

Filas, R. 2010. *Dwadzieścial at przemian polskich mediów masowych (1989-2009) w ujęciu periodycznym*, Zeszyty Prasoznawcze, pp. 27–54

Foucault, M. 1977. *Discipline and Punish*. London: GamsonHerzog.

Geertz, C. 1993. *The Interpretation of Cultures*. London: Fontana Press.

Gigantyczny magazyn na gaz powstanie pod Stargardem. 2012. “Szczecin Biznes.” http://szczecinbiznes.pl/informacje/Gigantyczny-magazyn-na-gaz-powstanie-pod-Stargardem-_1890?fbclid=IwAR1D77IDSrnZblXVrMcaHtmnIMMqMloLLV8W1Asep9VZ-elXWVhcW6l3yAE

Global CCS Institute (GCCSI). 2022. The Global Status of CCS: 2022, Australia.

Global CCS Institute (GCCSI). 2023. The Global Status of CCS: 2023, Australia.

Gordon, J.A., Balta-Ozkan, N., & Ali Nabavi, S. 2022. Beyond the triangle of renewable energy acceptance: The five dimensions of domestic hydrogen acceptance. *Applied Energy* 324: 119715

Gorzelak, G., & Jałowiecki, B. 1998. Koniunktura gospodarcza i mobilizacja społeczna w gminach w latach 1995-1997. w: Gorzelak G., & Jałowiecki, B. (Eds.) *Koniunktura gospodarcza i mobilizacja spoleczna w gminach*, Warszawa: Europejski Instytut Rozwoju Regionalnego i Lokalnego UW.

Gross, M. 2007. The unknown in process: Dynamic connections of ignorance, non-knowledge and related concepts. *Current Sociology*, 55(5): 742–759.

Guevara, P. 2024. *Understanding Energy Security*. https://safetyculture.com/topics/energy-security/

Gunderson, R., Stuart, D., & Petersen, B. 2020. The fossil fuel industry's framing of carbon capture and storage: Faith in innovation, value instrumentalization, and status quo maintenance. *Journal of Cleaner Production*, 252(119767): 1–9.

Haas., T., Herberg, J., & Löw-Beer, D. 2022 From carbon democracy to post-fossil capitalism? The German coal phase-out as a crossroads of sustainability politics. *Sustainability: Science, Practice and Policy*, 18(1): 384–399. https://doi.org/10.1080/15487733.2022.2069542

Habermas, J. 1991. The public sphere. In Mukerji, C., & Schudson, M. (Eds.), *Rethinking Popular Culture. Contemporary Perspectives in Cultural Studies*. Berkeley/Los Angeles: University of California Press, pp. 398–404.

Ha-Duong, M., Campos, A.S., & Nadai, A. 2009. A survey on the public perception of CCS in France. *International Journal of Greenhouse Gas Control*, 3: 633–640.

International Energy Agency (IEA). 2022. World Energy Outlook 2022, Paris, Licence: CC BY 4.0 (report). https://www.iea.org/reports/world-energy-outlook-2022.

IPCC Sixth Assessment Report. 2022. Climate Change 2022: Impacts, adaptation and vulnerability, Geneva.

IPSOS. 2023. IPSOS Global Trustworthiness Index. https://www.ipsos.com/en/politicians-least-trusted-profession-while-doctors-most-trustworthy

Jones, Ch. 2024. The role of CCS in the EU – thoughts on possible Road Map for the next Commission, RSC Policy Paper 2024/02. https://cadmus.eui.eu/bitstream/handle/1814/77197/RSC_PP_2024_02.pdf?sequence=1&isAllowed=y

Kaiser, M., Zimmer, R., Brunsting, S., Mastop, J., & Pol, M. 2014. Development of CCS projects in Poland. How to communicate with the local public? *Energy Procedia*, 51: 267–273. https://doi.org/10.1016/j.egypro.2014.07.031

Karimi, F., & Rodi, M. (Eds.). 2022. Energy transition in the Baltic Sea Region. *Understanding Stakeholders Engagement and Community Acceptance*. Oxon & New York: Routledge.

Langhelle, O., Sareen, S., & Silvester, B.R. 2024. From CCS to CCU and CCUS – The pitfalls of utilization and storage. In: Stokke, Ø. (Ed.), *Making* CO_2 *a Resource*. Abingdon: Routledge, pp. 127–142.

Li, Q., Liu, L.-C., Chen, Z.-A., Zhang, X., Jia, L., & Liu, G. 2014. A survey of public perception of CCUS in China. *Energy Procedia* 63: 7019–7023.

Lisek-Michalska, J. 2013. *Badania fokusowe. Problemy metodologiczne i etyczne*. Wydawnictwo UŁ: Łódź.

L'Orange Seigo, S., Dohle, S., & Siegrist, M. 2014. Public perception of carbon capture and storage (CCS): A review. *Renewable and Sustainable Energy Reviews*, 38: 848–863.

Nicolas, C., Tchung-Ming, S., Bahn, O., & Delage, E. 2021 Robust enough? Exploring temperature-constrained energy transition pathways under climate uncertainty. *Energies*, 14: 8595. https://doi.org/10.3390/en14248595

Nielsen, J.A.E., Stavrianakis, K., & Morrison, Z. 2022. Community acceptance and social impacts of carbon capture, utilization and storage projects: A systematic meta-narrative literature review. *PLOS ONE*, 17(8): e0272409. https://doi.org/10.1371/journal.pone.0272409

Nowotny, H., Scott, P., & Gibbons, M. 2001. *Re-Thinking Science. Knowledge and the Public in an Age of Uncertainty*. Cambridge: Polity Press.

Otto, D., & Gross, M. 2021. Stuck on coal and persuasion? A critical review of carbon capture and storage communication. *Energy Research & Social Science*, 82: 102306.

Otto, D., Pfeiffer, M., de Brito, M.M., & Gross, M. 2022. Fixed amidst change: 20 years of media coverage on carbon capture and storage in Germany. *Sustainability*, 4: 7342. https://doi.org/10.3390/su14127342

Otto, D., Sprenkeling, M., Peuchen, R., Nordø, Å.D., Mendrinos, D., Karytsas, S., Veland, S., Polyzou, O., Lien, M., Heggelund, Y., et al. 2022. On the organisation of translation – An inter- and transdisciplinary approach to developing design options for CO_2 storage monitoring systems. *Energies*, 15, 5678. https://doi.org/10.3390/en15155678

Pasman, H., Sripaul E., Khan F., & Fabiano B. 2023. Energy transition technology comes with new process safety challenges and risks, *Process Safety and Environmental Protection*, 177: 765–794. https://doi.org/10.1016/j.psep.2023.07.036.

Pidgeon, N., Corner, A., Parkhill, K., Spence, A., Butler, K., & Poortinga, W. 2012. Exploring early public responses to geoengineering. *Philosophical Transactions of the Royal Society A*, 370: 4176–4196. https://doi.org/10.1098/rsta.2012.0099

Peters, J.D. 1997. Seeing bifocally: Media, place, culture. In: Gupta A., & Ferguson J. (Eds.), *Culture, Power, Place*. Abington: Routledge, pp. 75–92.

Podemski, K. 2013. *Badania polskiego dyskursu publicznego II RP, PRL, i III RP. Przegląd zagadnień*, Kultura I społeczeństwo, pp. 27–66

Quang, D., Milani, D., & Zahra, M., 2023. A review of potential routes to zero and negative emission technologies via the integration of renewable energies with CO_2 capture processes, *International Journal of Greenhouse Gas Control*, 124: 103862. https://doi.org/10.1016/j.ijggc.2023.103862

Rothwell, E., Anderson, R., & Botkin, J.R. 2016. *Deliberative discussion focus groups. Qualitative Health Research*, 26(6): 734–740. https://doi.org/10.1177/1049732315591150

Røyrvik, J., Schei Olsen, M., & Aasen, T.M.B. 2012. Political rationality and CCS discourse. *Energy Procedia*, 23: 284–295

Rychlicki, S., Kosowski, P., Wartak, J., & Solecki, M. 2015. Social Acceptance for CO_2 or and CCS Projects Based on Survey Conducted in Southeastern Poland. *AGH Drilling Oil Gas* 32: 759–773.

Sadik-Zada, E.R., & Gatto, A. 2023. Civic engagement and energy transition in the nordic-baltic sea region: Parametric and nonparametric inquiries. *Socio-Economic Planning Sciences*, Volume 87: 101347. https://doi.org/10.1016/j.seps.2022.101347

Slakaityte, V., Surwillo, I., & Villumsen, T. 2020. *Energy Security in the Baltic Sea Region*. Copenhagen: DIIS Danish Institute for International Studies.

Sitinjak, C., Ebennezer, S., & Ober, J. 2023. Exploring public attitudes and acceptance of CCUS technologies in JABODETABEK: A cross-sectional study. *Energies*, 16: 4026. https://doi.org/10.3390/en16104026

Statistical Yearbook of Zachodniopomorskie Voivodship, Szczecin 2023

Statistics Poland. *Information society in Poland in 2023*, Warszawa-Szczecin 2023

Strojny, M., Gładysz, P., & Nowak, W. 2024. Preliminary comprehensive assessment of CO_2 utilization versus CO_2 storage in Poland, *Sustainable Energy Technologies and Assessments*, Vol.66: 103817. https://doi.org/10.1016/j.seta.2024.103817

Svartdal, I.H., & Kristoffersen, B. 2023. Why in my backyard (WIMBY): Forging the link to community futures when energy transition projects are met with indifference. *Energy Research & Social Science* 101: 103138. https://doi.org/10.1016/j.erss.2023.103138

Tarkowski, R., Luboń, K., & Tarkowski, S. 2014. Postrzeganie zmian klimatu oraz CCS – wyniki badań ankietowych społeczności lokalnej na przykładzie okolic Tarnowa. *Polityka Energetyczna – Energy Policy Journal*, 17: 85–98.

Tcvetkov, P., Cherepovitsyn, A., & Fedoseev, S. 2019. Public perception of carbon capture and storage: A state-of-the-art overview. *Helyon* 5: 1–28. https://doi.org/10.1016/j.heliyon.2019.e02845

Thomas, S. 2008. Enhanced oil recovery – An overview, oil & gas science and technology. *Revue d'IFP Energies nouvelles*, 63(1): 9–19.

Touraine, A. 2011. *Myśleć inaczej*. Warszawa: Państwowy Instytut Wydawniczy.

Tsur, Y. 2025. The diverse impacts of democracy on greenhouse gas emissions. *Ecological Economics*, 227: 108411. https://doi.org/10.1016/j.ecolecon.2024.108411.

United Nations (UN). 1972. Conference on the Human Environment, 5–16 June 1972, Stockholm.

United Nations (UN). 2015. Sustainable Development Summit, 25–27 September 2015, New York.

United Nations (UN). 2015. The Paris Agreement, 12 December 2015, Paris.

United Nations (UN). 2022. Economic Commission for Europe, 2022. Carbon Capture, Use and Storage (CCUS). Technology Brief, New York.

Upham, P., Oltra, C., & Boso, À. 2015. Towards a cross-paradigmatic framework of the social acceptance of energy systems. *Energy Research & Social Science*, 8: 100–112. https://doi.org/10.1016/j.erss.2015.05.003

Vercelli, S., Anderlucci, J., Memoli, R., Battisti, N., Mabon, L., & Lombardi, S. 2013. Informing people about CCS: A review of social research studies. *Energy Procedia*, 37: 7464–7473.

Wagner, O., & Götz, T. 2021. Presentation of the 5Ds in energy policy: A policy paper to show how Germany can regain its role as a Pioneer in energy policy. *Energies*, 14: 6799. https://doi.org/10.3390/en14206799

Wagner, W., Kronberger, N., & Seifert, F. 2002. *Collective symbolic coping with new technology: Knowledge, images and public discourse*. British Journal of Social Psychology, 41/2002, 323–343

Wassermann, S., Schulz, M., & Scheer, D. 2011. Linking public acceptance with expert knowledge on CO_2 storage: Outcomes of a delphi approach. *Energy Procedia*, 4: 6353–6359

Wind Power. 2024 Poland wind farm files. https://www.thewindpower.net/country_maps_en_27_poland.php

Wüstenhagen, R., Wolsink, M., & Bürer, M.J. 2007. Social acceptance of renewable energy innovation: An introduction to the concept. *Energy Policy*, 35: 2683–2691. https://doi.org/10.1016/j.enpol.2006.12.001

Xenias, D., & Whitmarsh, L. 2018. Carbon capture and storage (CCS) experts' attitudes to and experience with public engagement. *International Journal of Greenhouse Gas Control*, 78: 103–116.

Zachodniopomorskie Voivodship. Subregions, Powiats, Gminas, Szczecin: Statistical Office in Szczecin 2019.

Zarycki, T. 2002. *Region jako kontekst zachowań politycznych*, Warszawa: Scholar.

Internet and electronic media sources:

Local DataBase GUS https://bdl.stat.gov.pl/bdl/dane/podgrup/tablica

Samorząd. 2025. https://samorzad.pap.pl/kategoria/aktualnosci/juz-139-miast-na-liscie-tracacych-funkcje-spoleczno-gospodarcze-tak-walcza-z

https://sejmsenat2019.pkw.gov.pl/sejmsenat2019/pl/wyniki/sejm/okr/41 https://sejmsenat2019.pkw.gov.pl/sejmsenat2019/pl/wyniki/sejm/okr/40

https://ekrs.ms.gov.pl/web/wyszukiwarka-krs/strona-glowna/ (Krajowy Rejestr Sądowy)

https://widoczni.com/blog/najpopularniejsze-wyszukiwarki/, https://green-fields.pl/ranking-wyszukiwarek-2020-2021-2022/; https://www.artefakt.pl/blog/seo/wyszukiwarka-google-nadal-bezkonkurencyjna-2022

National Energy and Climate Plan for 2021-2030. Assumptions and goals as well as policies and activities, Ministry of State Assets. https://www.gov.pl/web/aktywa-panstwowe/krajowy-plan-na-rzecz-energii-i-klimatu-na-lata-2021-2030-przekazany-do-ke

National Strategy of Regional Development 2030. Ministry of Development Funds and Regional Policy. https://www.gov.pl/web/fundusze-regiony/krajowa-strategia-rozwoju-regionalnego

Poland's Energy Policy until 2040. Ministry of Climate and Environment. https://www.gov.pl/web/klimat/polityka-energetyczna-polski

Social Media. 2023. *GEMIUS-PBI-IAB Poland*. https://www.iab.org.pl/bez-kategorii/social-media-2023-pierwsza-edycja-raportu-gemius-pbi-i-iab-polska-juz-dostepna/

State Ecological Policy 2030. Ministry of Climate and Environment. https://www.gov.pl/web/srodowisko/polityka-ekologiczna-panstwa-2030–strategia-rozwoju-w-obszarze-srodowiska-i-gospodarki-wodnej?fbclid=IwAR0j6u3nGSlVYwlRhbgCdelXIVqyKCHp76yQnEwWvjnPohjEIlDJ9j7v5ps

Strategy for Development of West Pomeranian Province until 2030. West Pomeranian Self-Government. http://eregion.wzp.pl/sites/default/files/srwz_2030.pdf

Strategy for Responsible Development by 2020 (with the perspective by 2030). Ministry of Development Funds and Regional Policy. https://www.gov.pl/web/fundusze-regiony/informacje-o-strategii-na-rzecz-odpowiedzialnego-rozwoju

Strategy for Sustainable Development in Transport until 2030. Ministry of Infrastructure. https://www.gov.pl/web/infrastruktura/projekt-strategii-zrownowazonego-rozwoju-transportu-do-2030-roku2

Strategy for Sustainable Development of Rural Areas, Agriculture and Fisheries until 2030. Ministry of Agriculture and Rural Development. https://www.gov.pl/web/rolnictwo/strategia-zrownowazonego-rozwoju-wsi-rolnictwa-i-rybactwa-2030

WORLD ENERGY TRILEMMA. 2024. *Evolving with Resilience and Justice*. London: World Energy Council.

Index

Note: **Bold** indicates tables and *italics* indicates figures in the text.